HARCOURT

Math

Teacher's Resource Book

Grade 3

Orlando • Boston • Dallas • Chicago • San Diego
www.harcourtschool.com

Copyright © by Harcourt, Inc.

All rights reserved. No part of this publication may be reproduced or transmitted in any form or by any means, electronic or mechanical, including photocopy, recording, or any information storage and retrieval system.

Teachers using HARCOURT MATH may photocopy complete pages in sufficient quantities for classroom use only and not for resale.

HARCOURT and the Harcourt Logo are trademarks of Harcourt, Inc.

Printed in the United States of America

ISBN 0-15-320948-8

5 6 7 8 9 10   018   2004 2003

CONTENTS

PROBLEM SOLVING

Problem Solving Think Along (written)...TR1
Problem Solving Think Along (oral)..........TR2

NUMBER AND OPERATIONS

Numeral Cards (0–7) TR3
Numeral Cards (8–15) TR4
Number Lines (by ones)............................TR5
Number Lines (by tens
 and hundreds) TR6
Base-Ten Materials (15 tens,
 20 units, for primary) TR7
Base-Ten Materials (2 flats,
 5 longs, 20 units) TR8
Base-Ten Materials
 (thousands cube)..................................TR9
Place-Value Charts (ones period) TR10
Place-Value Charts
 (to thousands period)........................TR11
Regrouping Workmat
 (for chip-trading—ones,
 tens, hundreds)....................................TR12
Place-Value Charts
 (for +, −, ×, hundreds,
 tens, ones)..TR13
Place-Value Workmat 5
 (hundreds, tens, ones) TR14
Hundred Chart TR15
Number Lines (for primary levels)............TR16
Addition/Multiplication Table (to 10).....TR17
Addition/Multiplication Table (to 12) TR18
Multiplication Table (filled to 10)............TR19
Multiplication Table (filled to 12) TR20
Equal Groups Workmat 6............................TR21
Fraction Circles (whole to fifths) TR22
Fraction Circles (sixths to tenths) TR23
Fraction Strips TR24
Number Lines (fractions) TR25
Decimal Models (tenths)...........................TR26
Decimal Models (hundredths)...................TR27
Number Lines (decimals)..........................TR28

GEOMETRY

Triangle Dot Paper TR29
Square Dot Paper......................................TR30
Cube Pattern TR31
Cylinder Pattern..TR32
Cone Pattern TR33
Square Pyramid Pattern TR34
Rectangular Prism Pattern TR35
Triangular Prism Pattern...........................TR36
Plane Figures (square, rectangle,
 triangle, circle) TR37
Plane Figures (for tessellations)................TR38
Circles (different sizes)..............................TR39
Triangles (different sizes)..........................TR40
Rectangles and Squares
 (different sizes) TR41
Polygons: Quadrilaterals TR42
Pattern Block Patterns TR43
Tangram Pattern..TR44

TIME, MONEY, MEASUREMENT

Blank Calendar TR45
January through June Calendars TR46
Large Analog Clock TR47
Analog Clockfaces TR48
Digital Clock Model TR49
Coins (pennies nickels,
    (dimes, quarters) TR50
Coins and Bills (quarters,
    half dollars, $1 bills) TR51
Bills ($1 bills, $5 bills) TR52
Money Workmat 4 (pennies nickels,
    dimes, quarters) TR53
Rulers (inches and centimeters) TR54
Thermometers
    (Celsius and Fahrenheit) TR55

DATA, PROBABILITY, AND STATISTICS

1-Inch Grid Paper TR56
1-Centimeter Grid Paper TR57
Pictograph Pattern TR58
Bar Graph Pattern 1 ....TR59
Bar Graph Pattern 2 ....TR60
Grid of Quadrant 1 ....TR61
Tally Table TR62
Line Graph Pattern TR63
Number Cube Patterns (with
    numbers 1–6 and without
    numbers) TR64
Spinners (blank and 2-section) TR65
Spinners (3- and 4-section) TR66
Spinners (5- and 6-section) TR67
Spinners (7- and 8-section) TR68
Spinners (9- and 10-section) TR69

TEACHER'S EDITION PRACTICE GAMES

High Number TR70
Beat the Clock TR71
Around the Moon TR72
Fact Family TR73
Spin Me TR74
Right Angle Ramble TR75
Ready, Set, Measure! TR76
Color It In TR77
Division Toss TR78
Bingo Grid TR79
Award Certificates TR80–81

DAILY FACTS PRACTICE TR82–104

FACT CARDS

Multiplication Facts TR105–118
Division Facts TR119–131

VOCABULARY CARDS

Vocabulary Cards TR133–168

TEACHER'S RESOURCE BOOK

This section includes various types of resources for lessons in *Harcourt Math*.

Resources are provided for the following categories:

- ▶ **Problem Solving**
- ▶ **Number and Operations**
- ▶ **Money, Measurement**
- ▶ **Data, Probability, and Graphing**
- ▶ **Geometry**
- ▶ **Teacher's Edition Practice Games**
- ▶ **Daily Facts Practice**
- ▶ **Fact Cards**

Name _____

PROBLEM SOLVING THINK ALONG

Problem Solving

Understand

1. Retell the problem in your own words. _____
_____
_____
_____

2. List the information given. _____
_____

3. Restate the question as a fill-in-the-blank sentence. _____
_____

Plan

4. List one or more problem-solving strategies that you can use. _____
_____

5. Predict what your answer will be. _____
_____

Solve

6. Show how you solved the problem. _____
_____
_____
_____

7. Write your answer in a complete sentence. _____
_____

Check

8. Tell how you know your answer is reasonable. _____
_____

9. Describe another way you could have solved the problem. _____
_____

Problem Solving Think Along (written) — Teacher's Resource Book TR1

Name _____

Problem Solving Think Along

Understand

1. What is the problem about?
2. What information is given in the problem?
3. What is the question?

Plan

4. What problem-solving strategies might I try to help me solve the problem?
5. About what do I think my answer will be?

Solve

6. How can I solve the problem?
7. How can I state my answer in a complete sentence?

Check

8. How do I know whether my answer is reasonable?
9. How else might I have solved this problem?

Numeral Cards (0–7) Teacher's Resource Book **TR3**

12	8
13	9
14	10
15	11

TR4 Teacher's Resource Book

Numeral Cards (8–15)

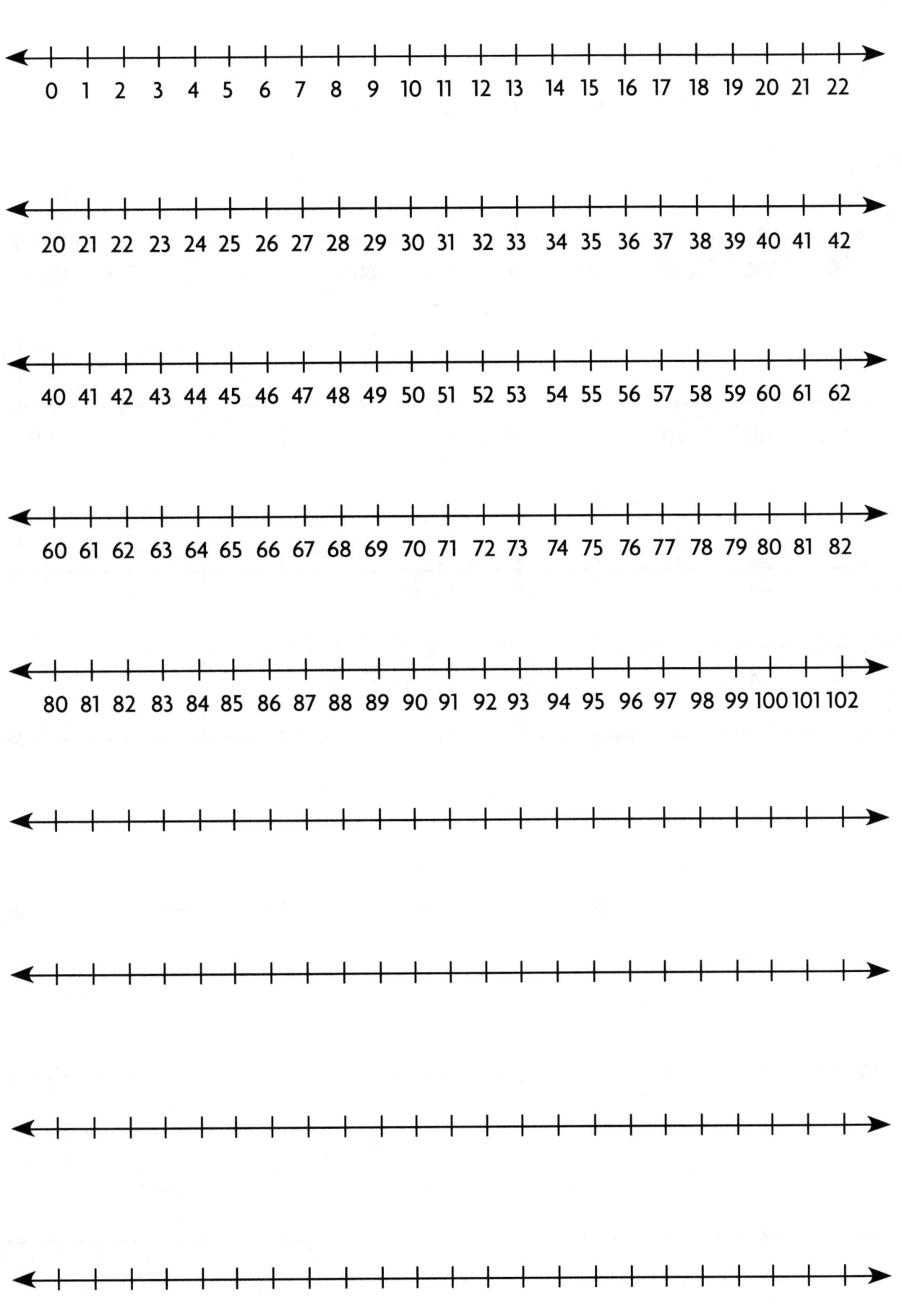

Number Lines · Teacher's Resource Book TR5

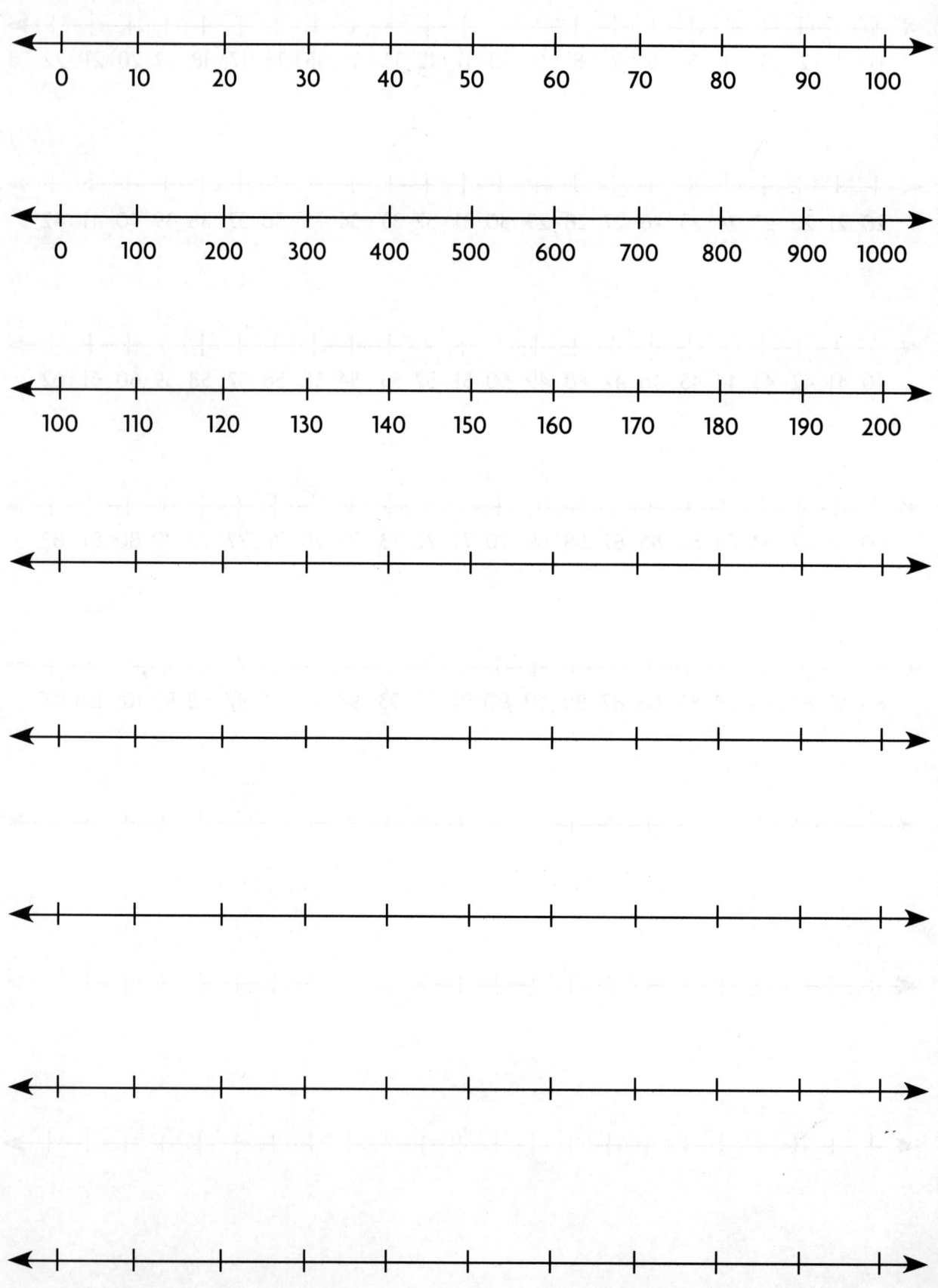

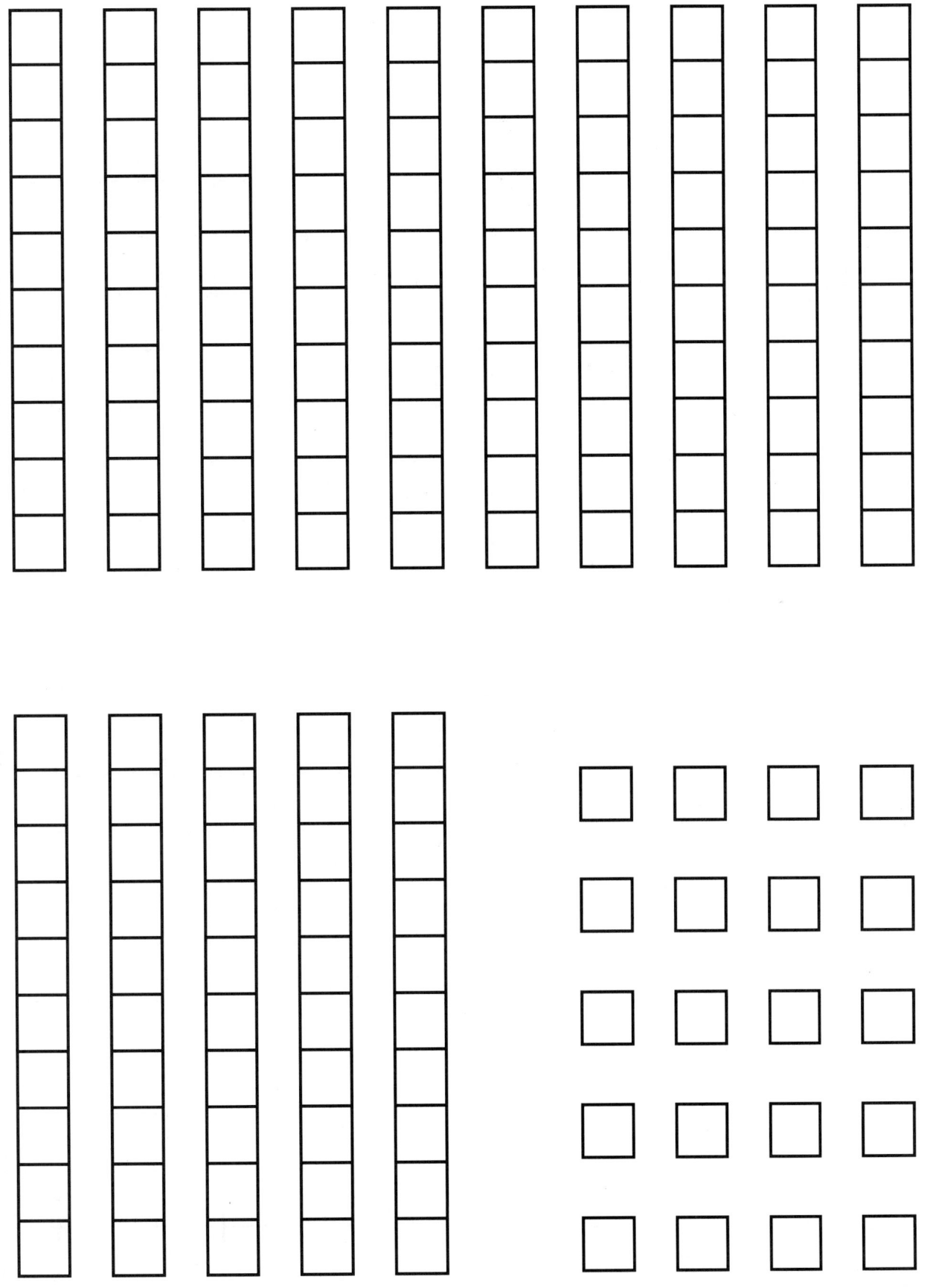

Base-Ten Materials · Teacher's Resource Book TR7

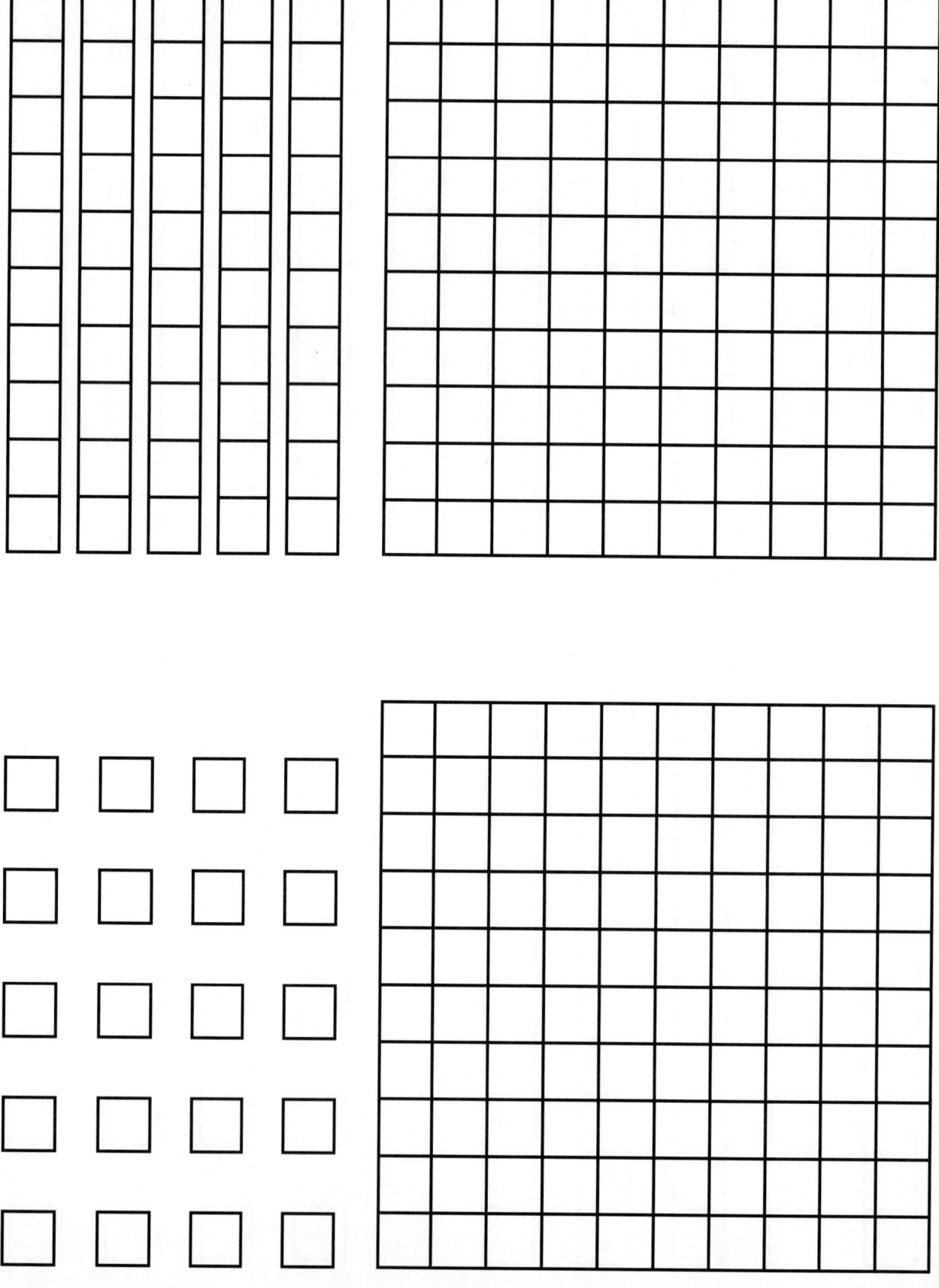

TR8 Teacher's Resource Book Base-Ten Materials

Base-Ten Materials

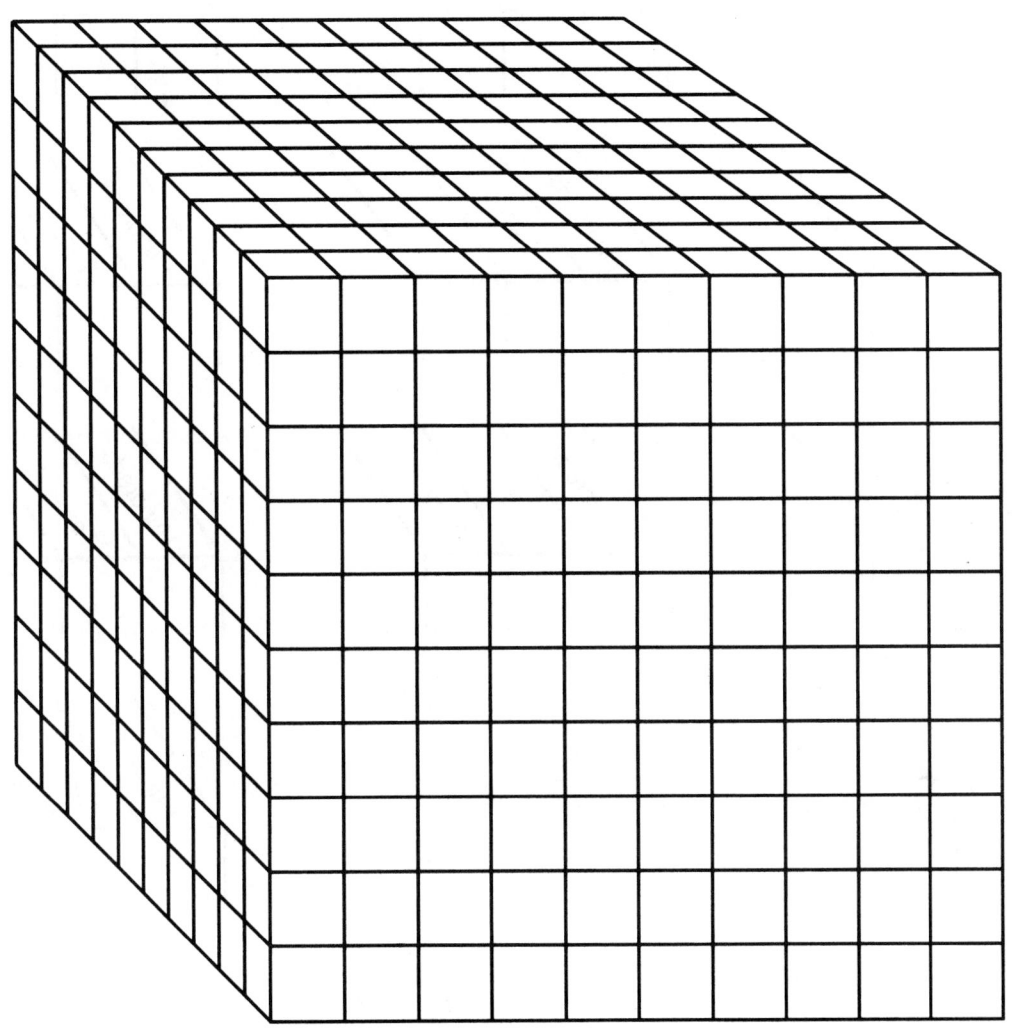

Teacher's Resource Book TR9

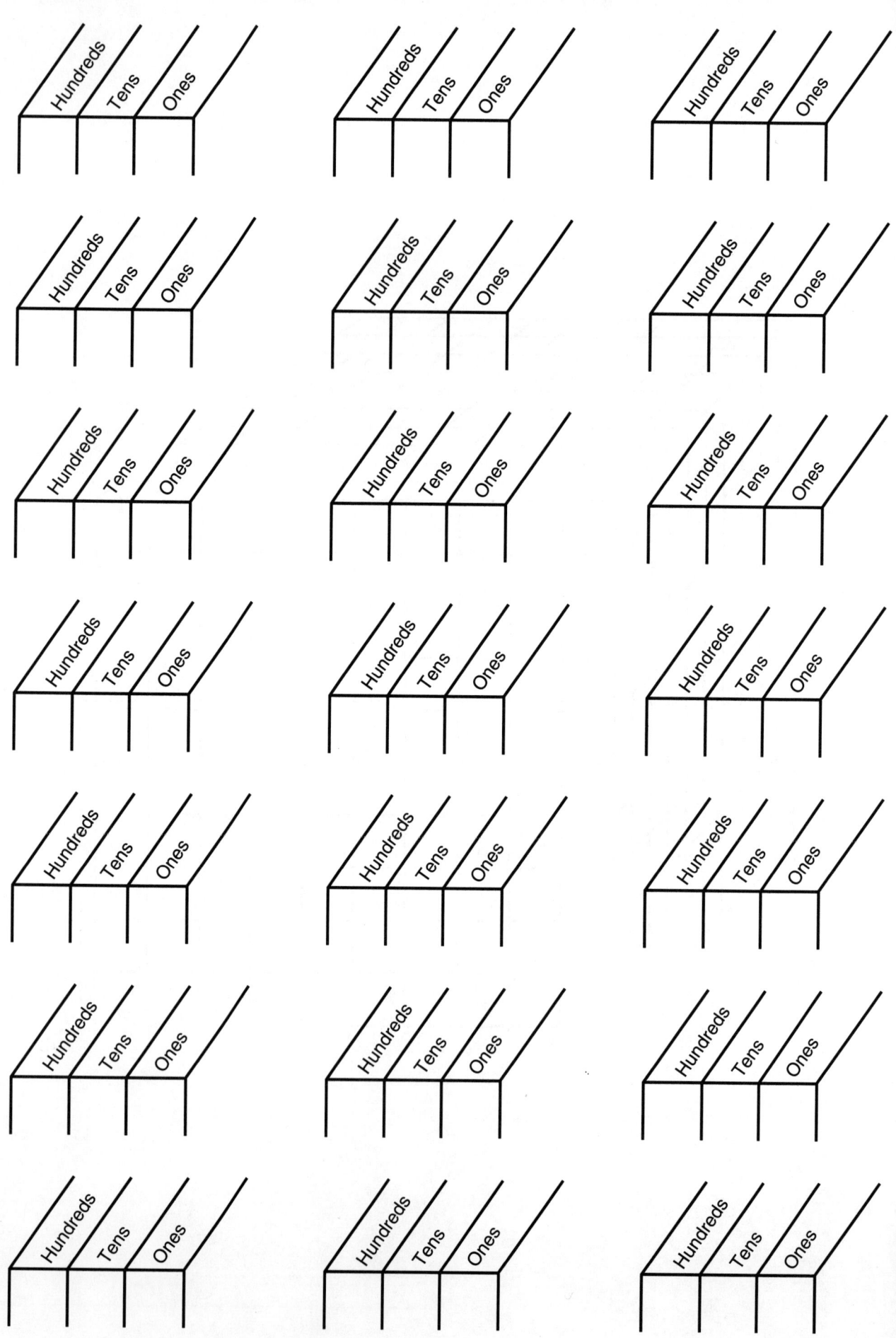

Place-Value Charts

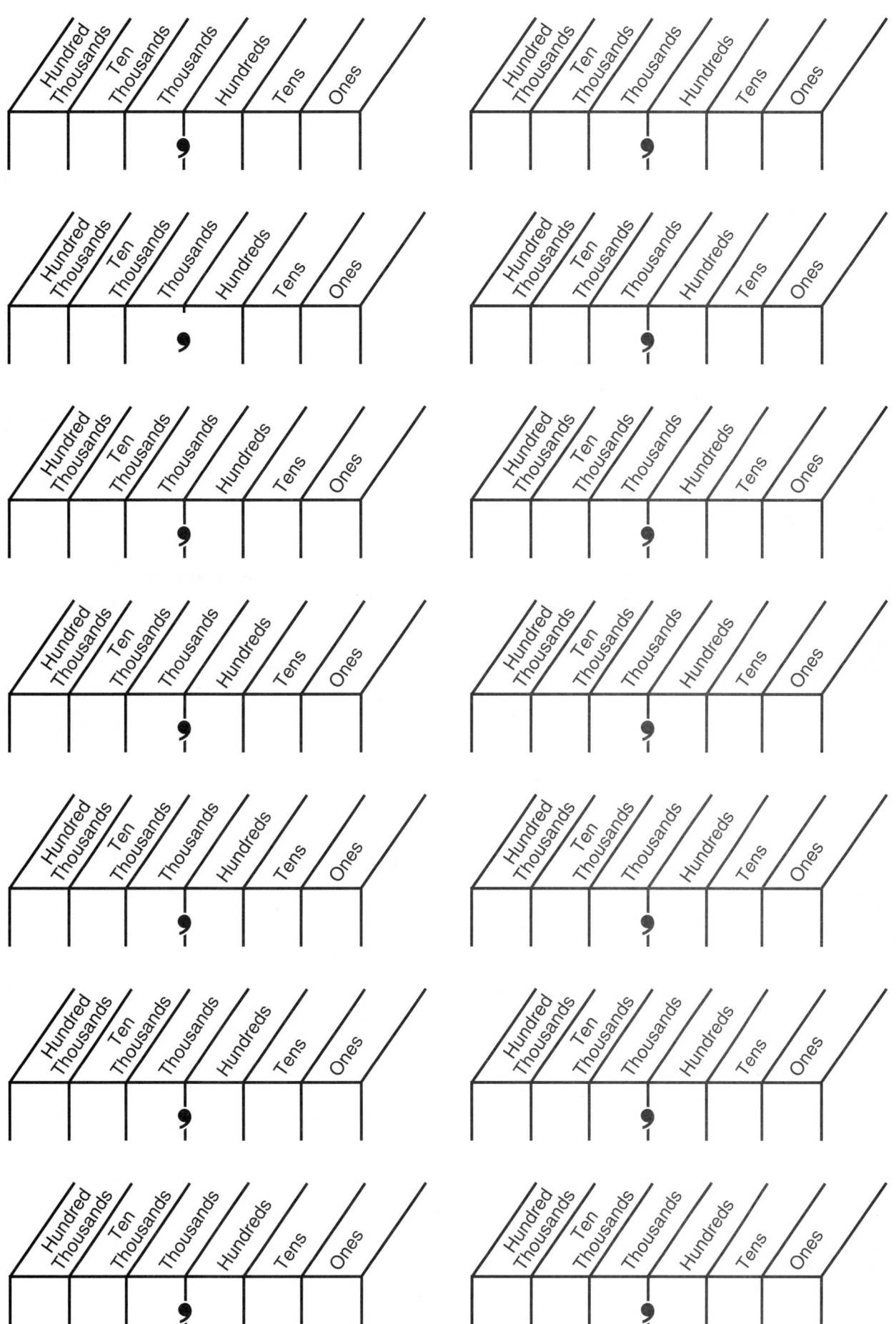

Place-Value Charts

Teacher's Resource Book TR11

TR12 **Teacher's Resource Book** Regrouping Workmat

| hundreds | tens | ones |

| hundreds | tens | ones |

| hundreds | tens | ones |

| hundreds | tens | ones |

| hundreds | tens | ones |

| hundreds | tens | ones |

| hundreds | tens | ones |

| hundreds | tens | ones |

Place-Value Charts

Workmat 5

| Hundreds | Tens | Ones |

TR14 Teacher's Resource Book — Place-Value Workmat

1	2	3	4	5	6	7	8	9	10
11	12	13	14	15	16	17	18	19	20
21	22	23	24	25	26	27	28	29	30
31	32	33	34	35	36	37	38	39	40
41	42	43	44	45	46	47	48	49	50
51	52	53	54	55	56	57	58	59	60
61	62	63	64	65	66	67	68	69	70
71	72	73	74	75	76	77	78	79	80
81	82	83	84	85	86	87	88	89	90
91	92	93	94	95	96	97	98	99	100

Hundred Chart

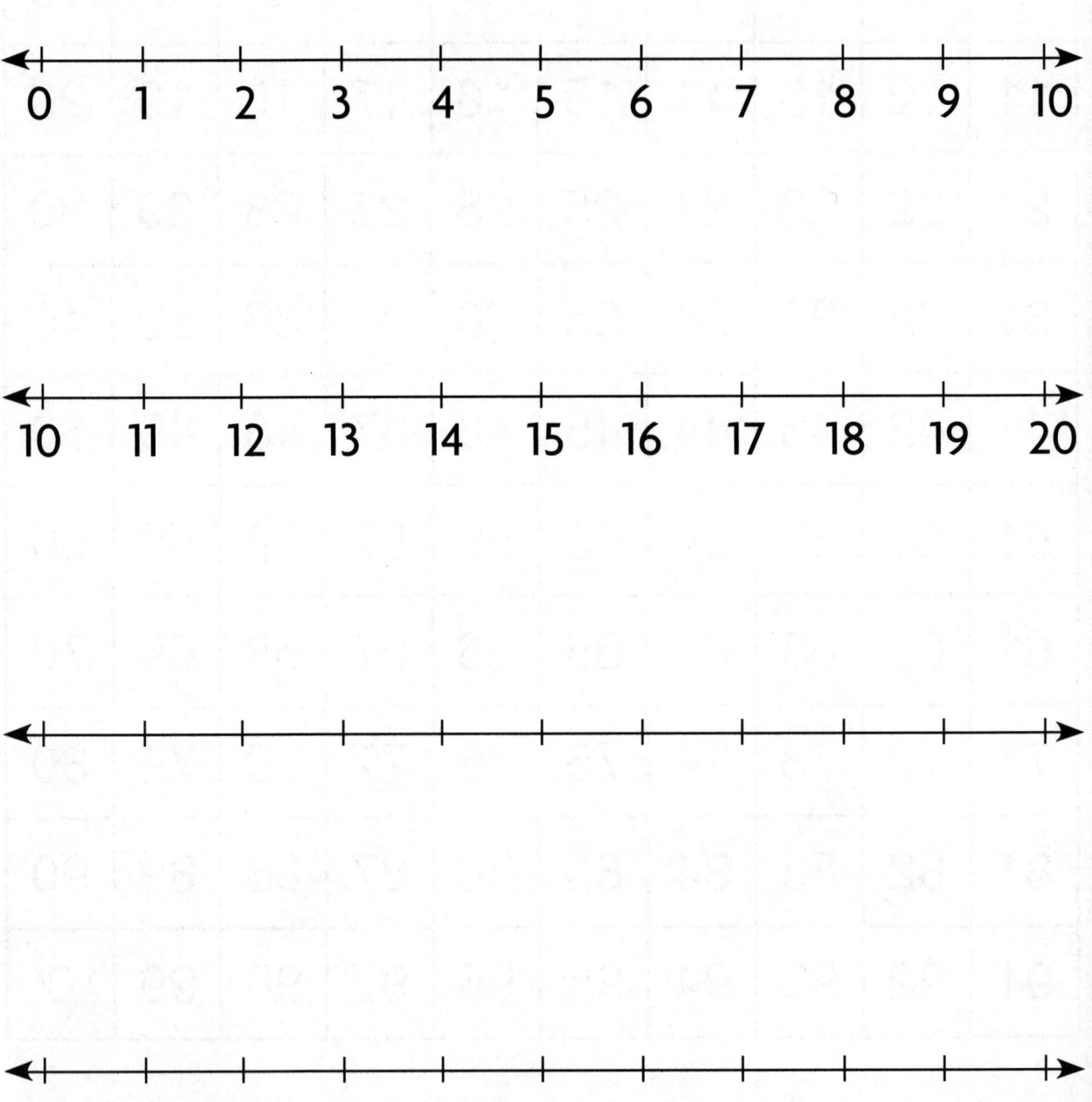

Addition/Multiplication Table

	0	1	2	3	4	5	6	7	8	9	10
0											
1											
2											
3											
4											
5											
6											
7											
8											
9											
10											

TR18 **Teacher's Resource Book** Addition/Multiplication Table

×	1	2	3	4	5	6	7	8	9	10
1	1	2	3	4	5	6	7	8	9	10
2	2	4	6	8	10	12	14	16	18	20
3	3	6	9	12	15	18	21	24	27	30
4	4	8	12	16	20	24	28	32	36	40
5	5	10	15	20	25	30	35	40	45	50
6	6	12	18	24	30	36	42	48	54	60
7	7	14	21	28	35	42	49	56	63	70
8	8	16	24	32	40	48	56	64	72	80
9	9	18	27	36	45	54	63	72	81	90
10	10	20	30	40	50	60	70	80	90	100

Multiplication Table (through 10)

×	1	2	3	4	5	6	7	8	9	10	11	12
1	1	2	3	4	5	6	7	8	9	10	11	12
2	2	4	6	8	10	12	14	16	18	20	22	24
3	3	6	9	12	15	18	21	24	27	30	33	36
4	4	8	12	16	20	24	28	32	36	40	44	48
5	5	10	15	20	25	30	35	40	45	50	55	60
6	6	12	18	24	30	36	42	48	54	60	66	72
7	7	14	21	28	35	42	49	56	63	70	77	84
8	8	16	24	32	40	48	56	64	72	80	88	96
9	9	18	27	36	45	54	63	72	81	90	99	108
10	10	20	30	40	50	60	70	80	90	100	110	120
11	11	22	33	44	55	66	77	88	99	110	121	132
12	12	24	36	48	60	72	84	96	108	120	132	144

Multiplication Table (through 12)

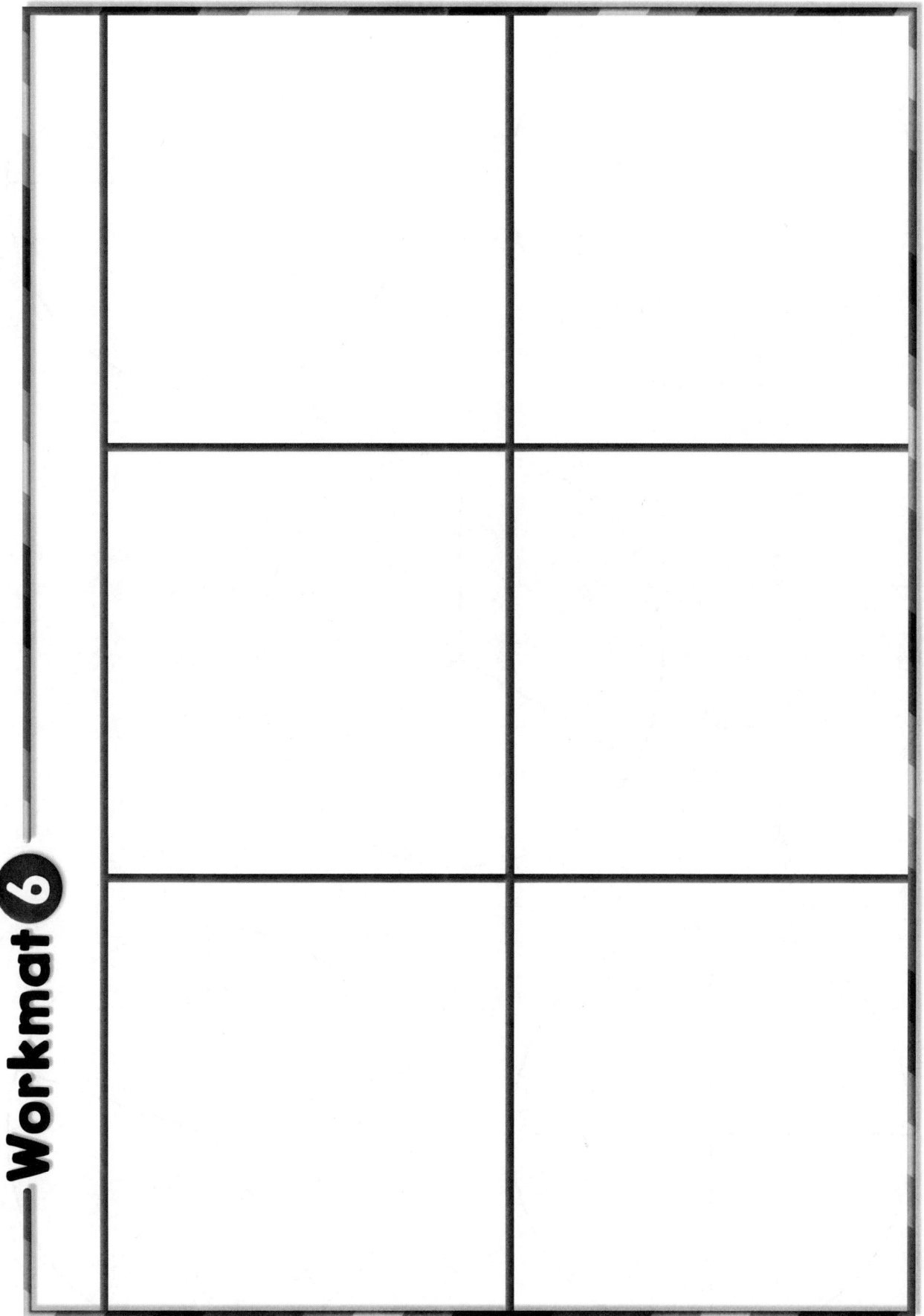

Equal Groups Workmat 6　　　　Teacher's Resource Book　　TR21

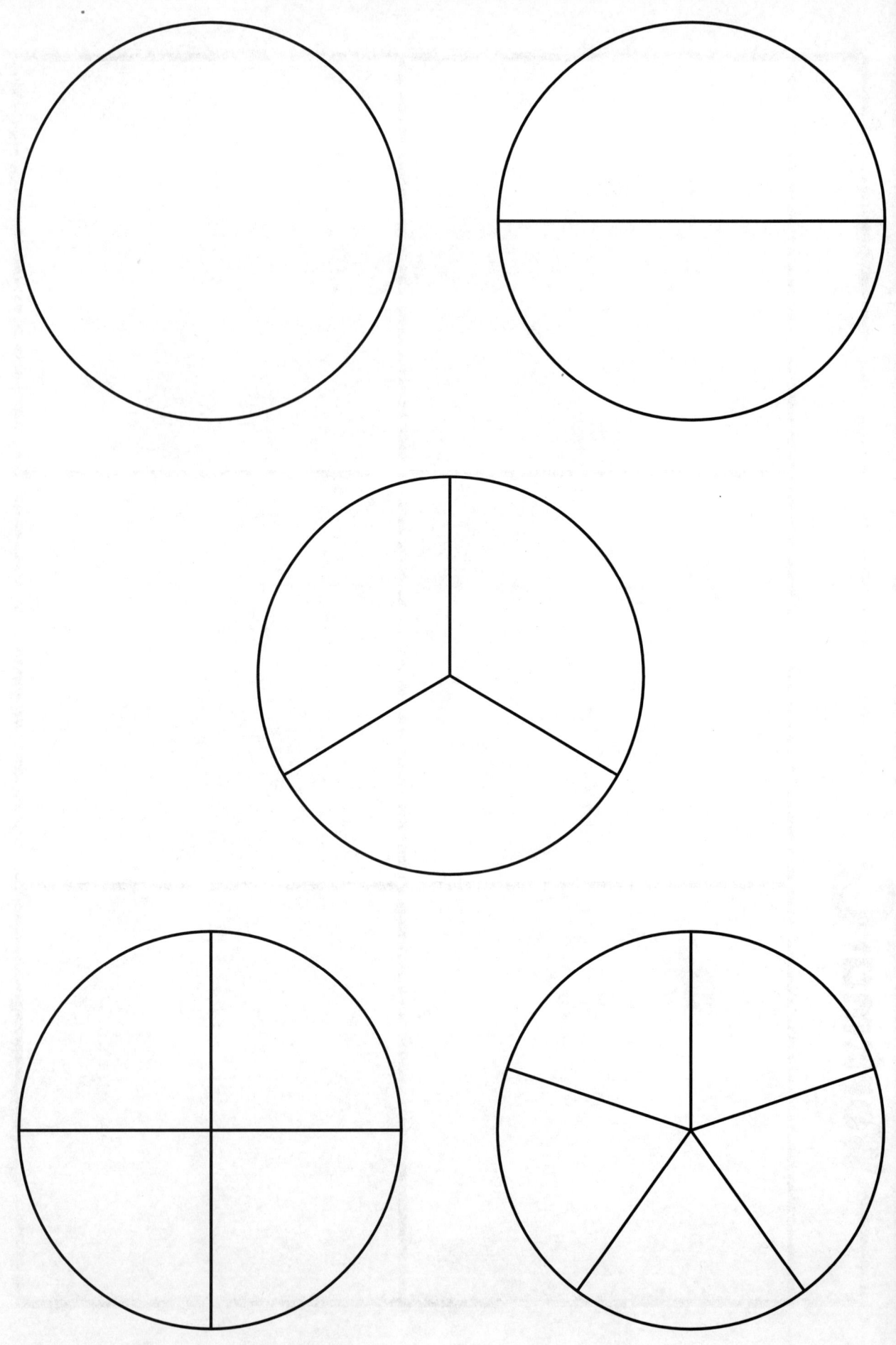

TR22 **Teacher's Resource Book** Fraction Circles

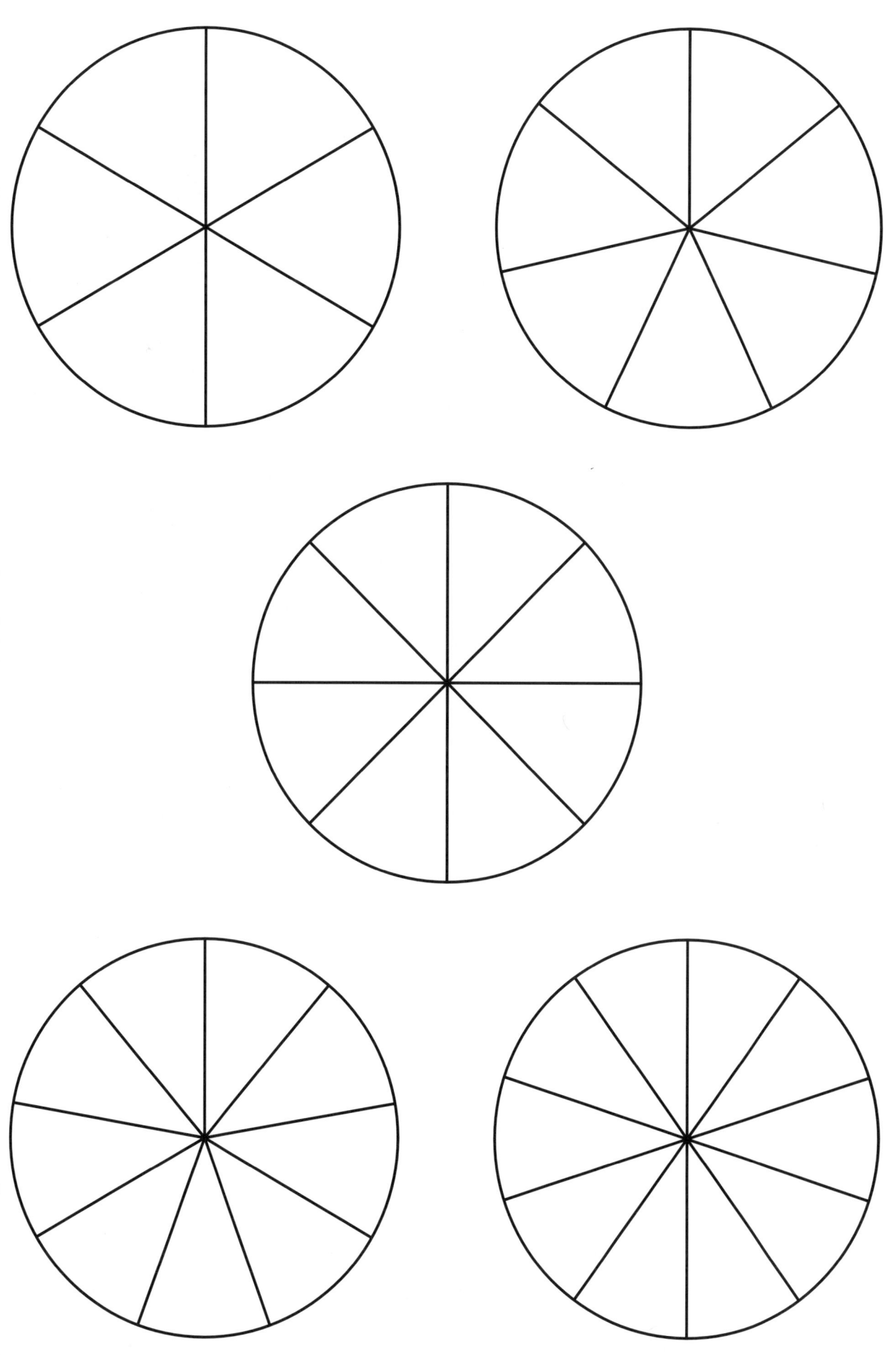

Fraction Circles   Teacher's Resource Book   TR23

Fraction Strips

| 1/12 | 1/12 | 1/12 | 1/12 | 1/12 | 1/12 | 1/12 | 1/12 | 1/12 | 1/12 | 1/12 | 1/12 |

| 1/11 | 1/11 | 1/11 | 1/11 | 1/11 | 1/11 | 1/11 | 1/11 | 1/11 | 1/11 | 1/11 |

| 1/10 | 1/10 | 1/10 | 1/10 | 1/10 | 1/10 | 1/10 | 1/10 | 1/10 | 1/10 |

| 1/9 | 1/9 | 1/9 | 1/9 | 1/9 | 1/9 | 1/9 | 1/9 | 1/9 |

| 1/8 | 1/8 | 1/8 | 1/8 | 1/8 | 1/8 | 1/8 | 1/8 |

| 1/7 | 1/7 | 1/7 | 1/7 | 1/7 | 1/7 | 1/7 |

| 1/6 | 1/6 | 1/6 | 1/6 | 1/6 | 1/6 |

| 1/5 | 1/5 | 1/5 | 1/5 | 1/5 |

| 1/4 | 1/4 | 1/4 | 1/4 |

| 1/3 | 1/3 | 1/3 |

| 1/2 | 1/2 |

| 1 |

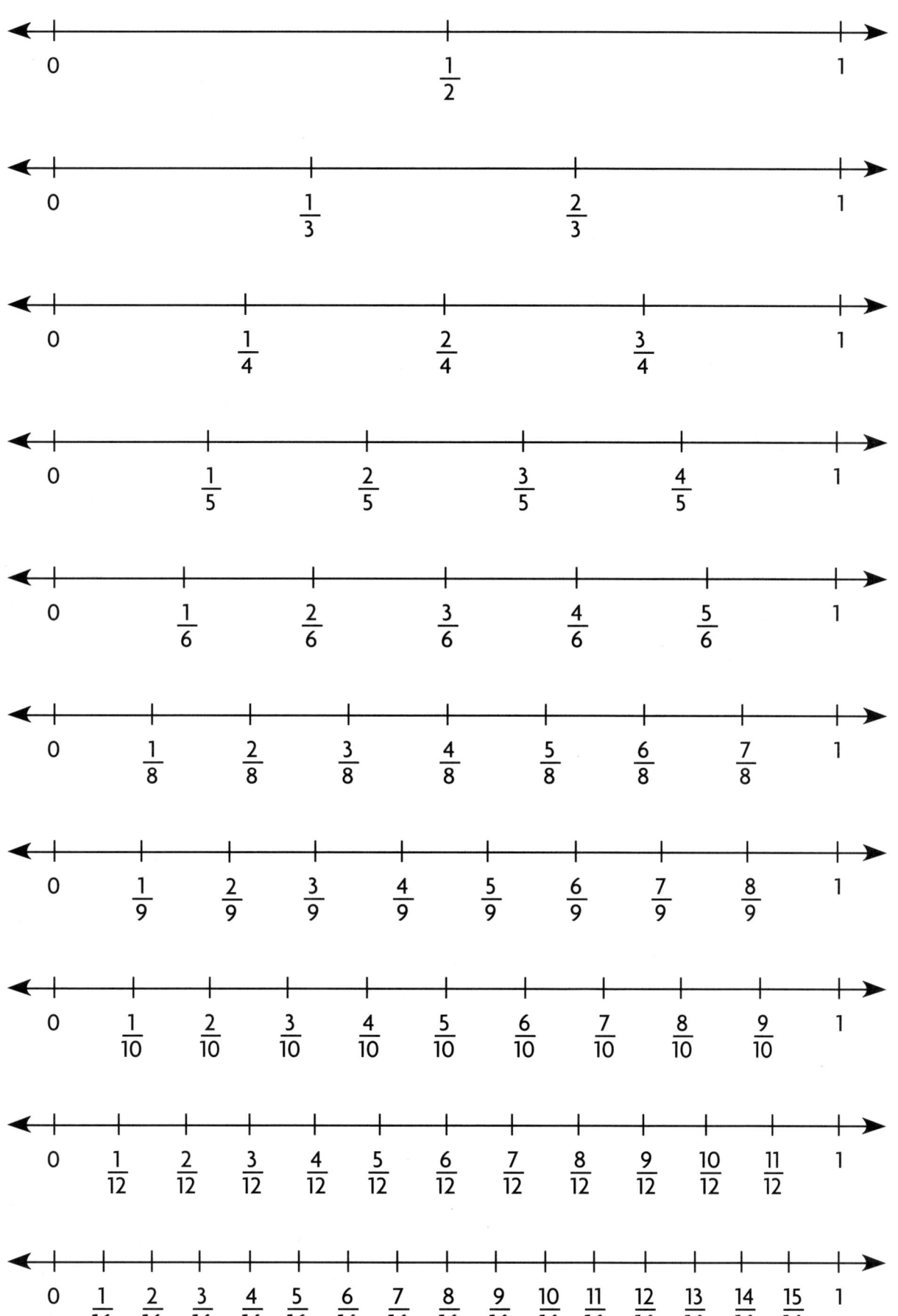

Number Lines

TR26 Teacher's Resource Book

Decimal Models

Decimal Models

Teacher's Resource Book TR27

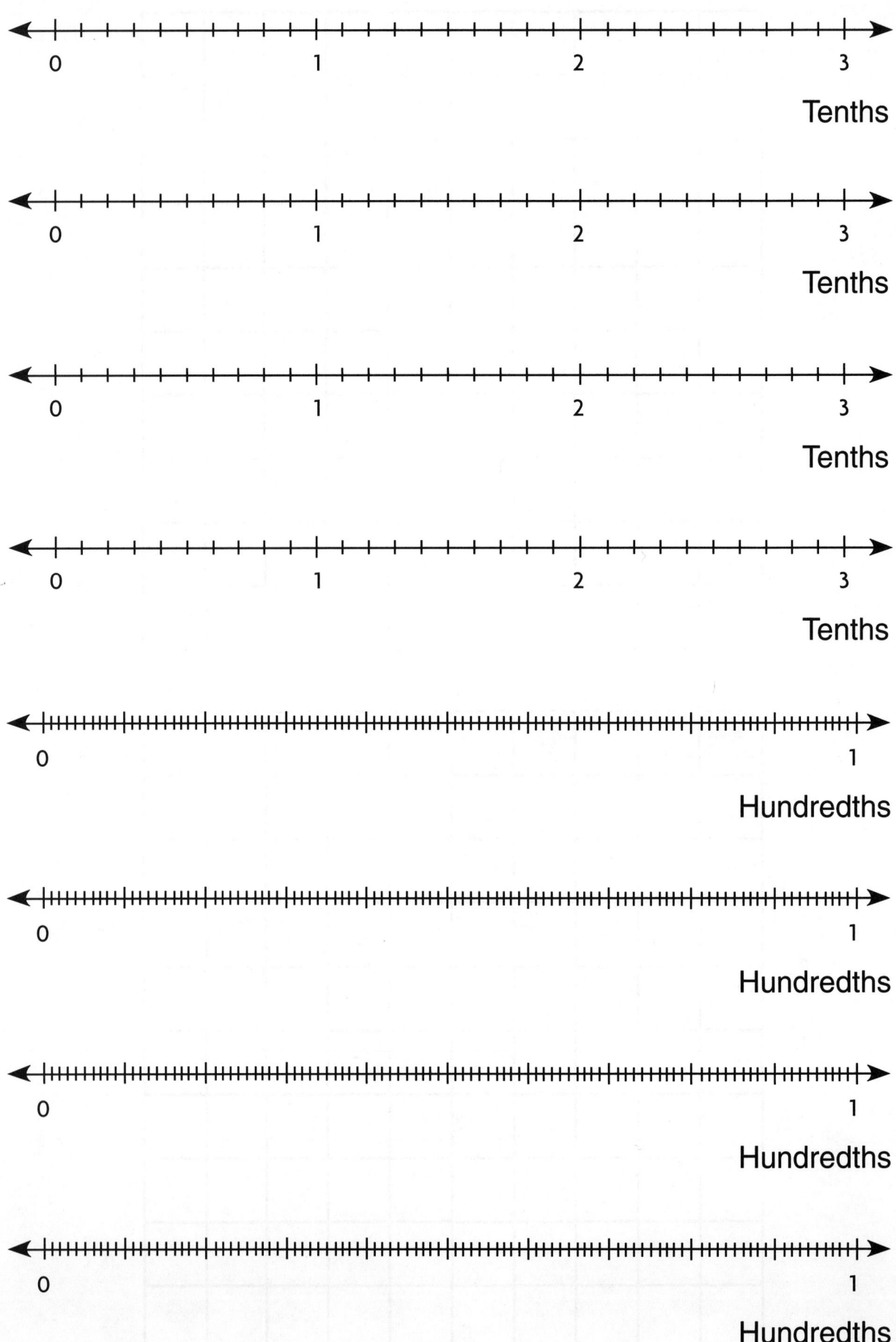

TR28 Teacher's Resource Book — Number Lines

Triangle Dot Paper   Teacher's Resource Book   **TR29**

TR30 Teacher's Resource Book

Square Dot Paper

Cube Pattern

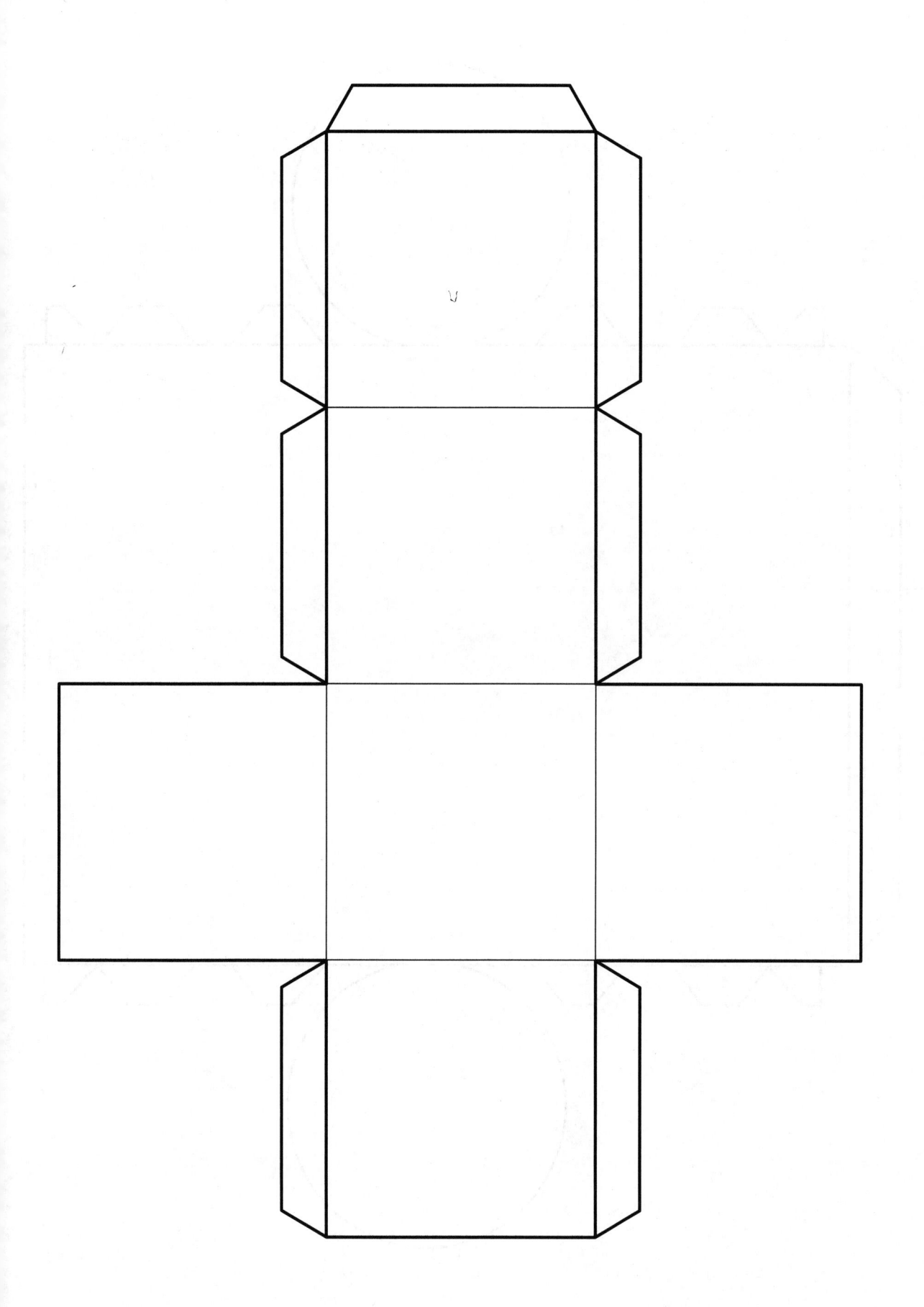

Teacher's Resource Book TR31

TR32 **Teacher's Resource Book**   Cylinder Pattern

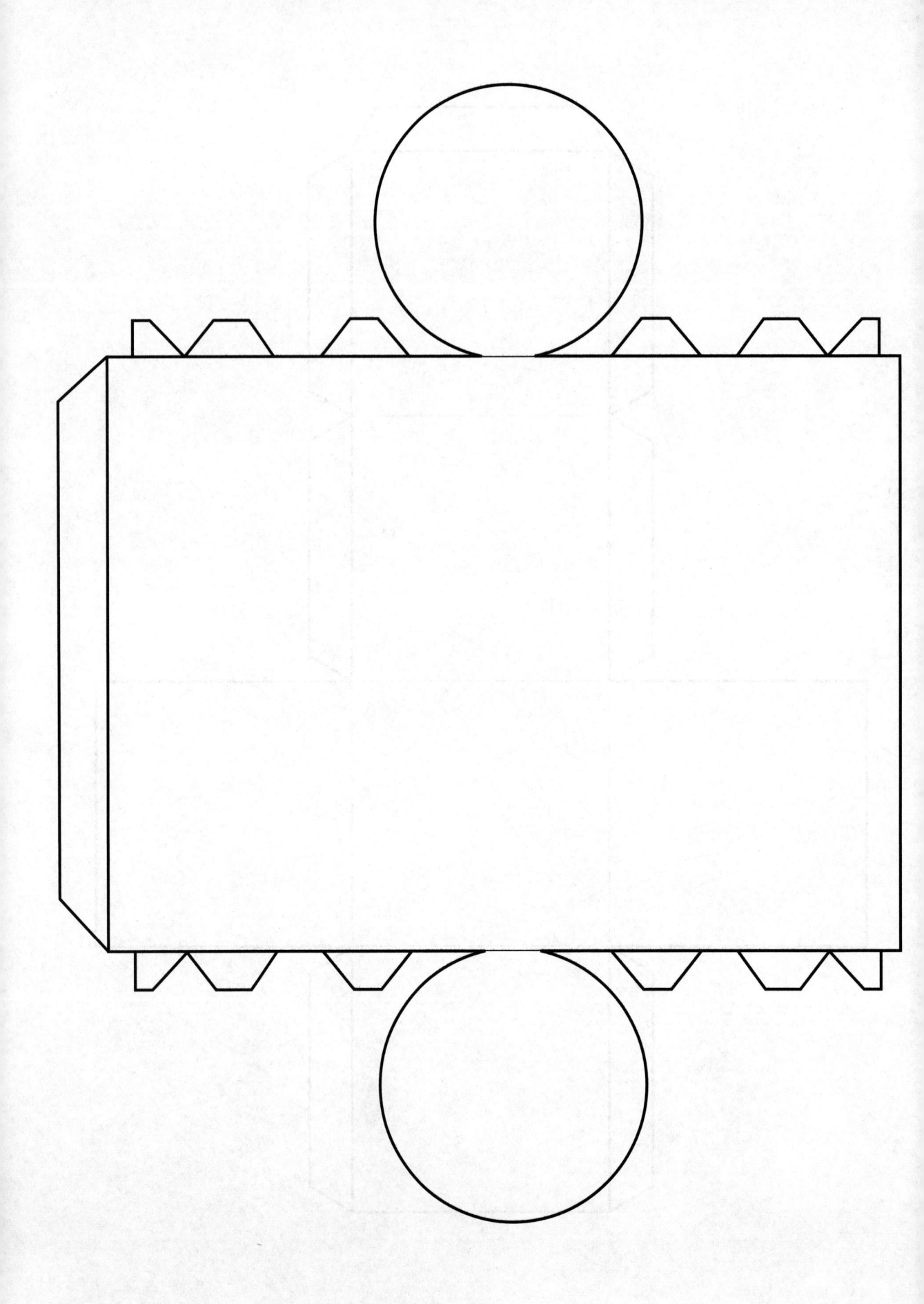

Cone Pattern

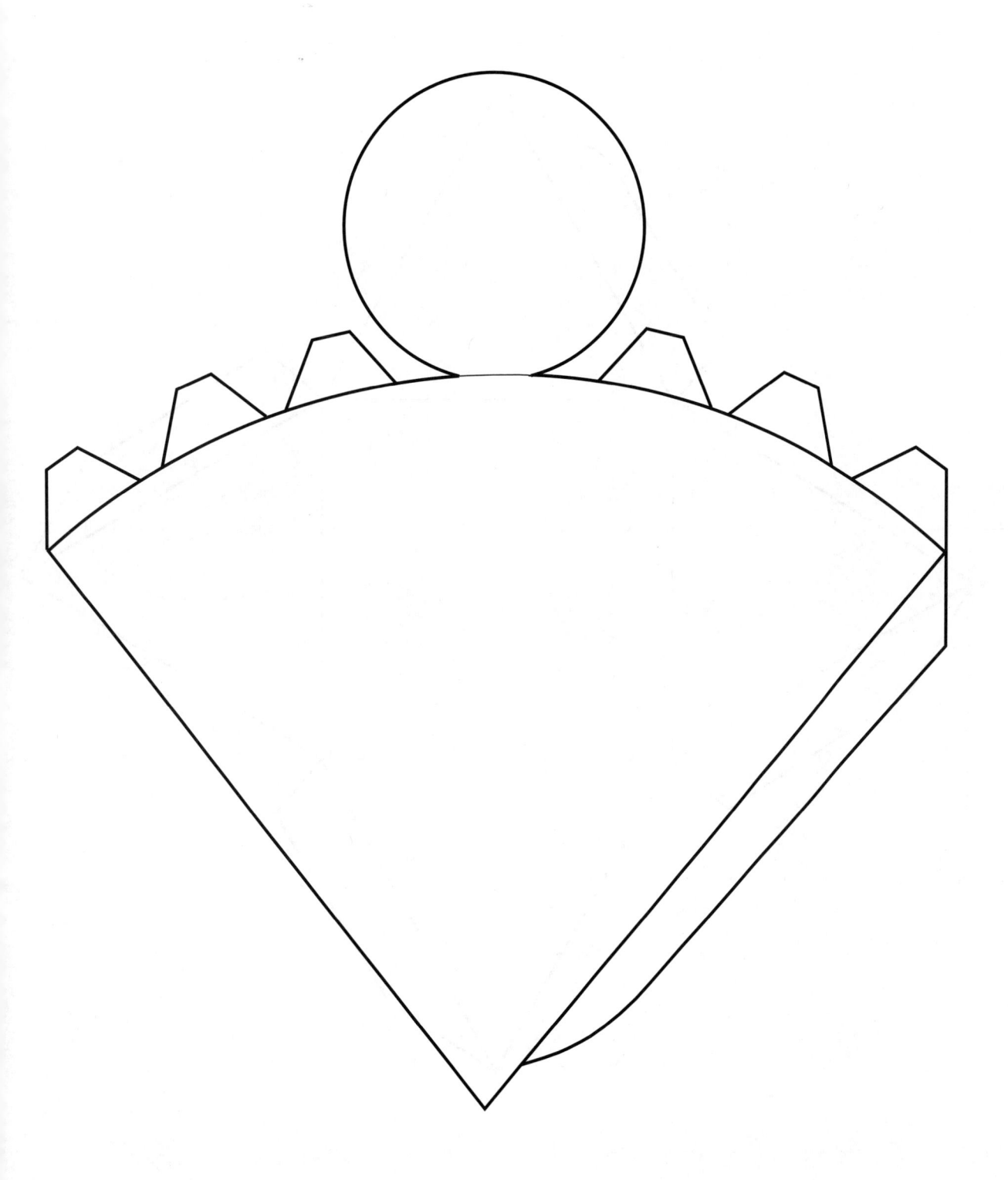

Teacher's Resource Book **TR33**

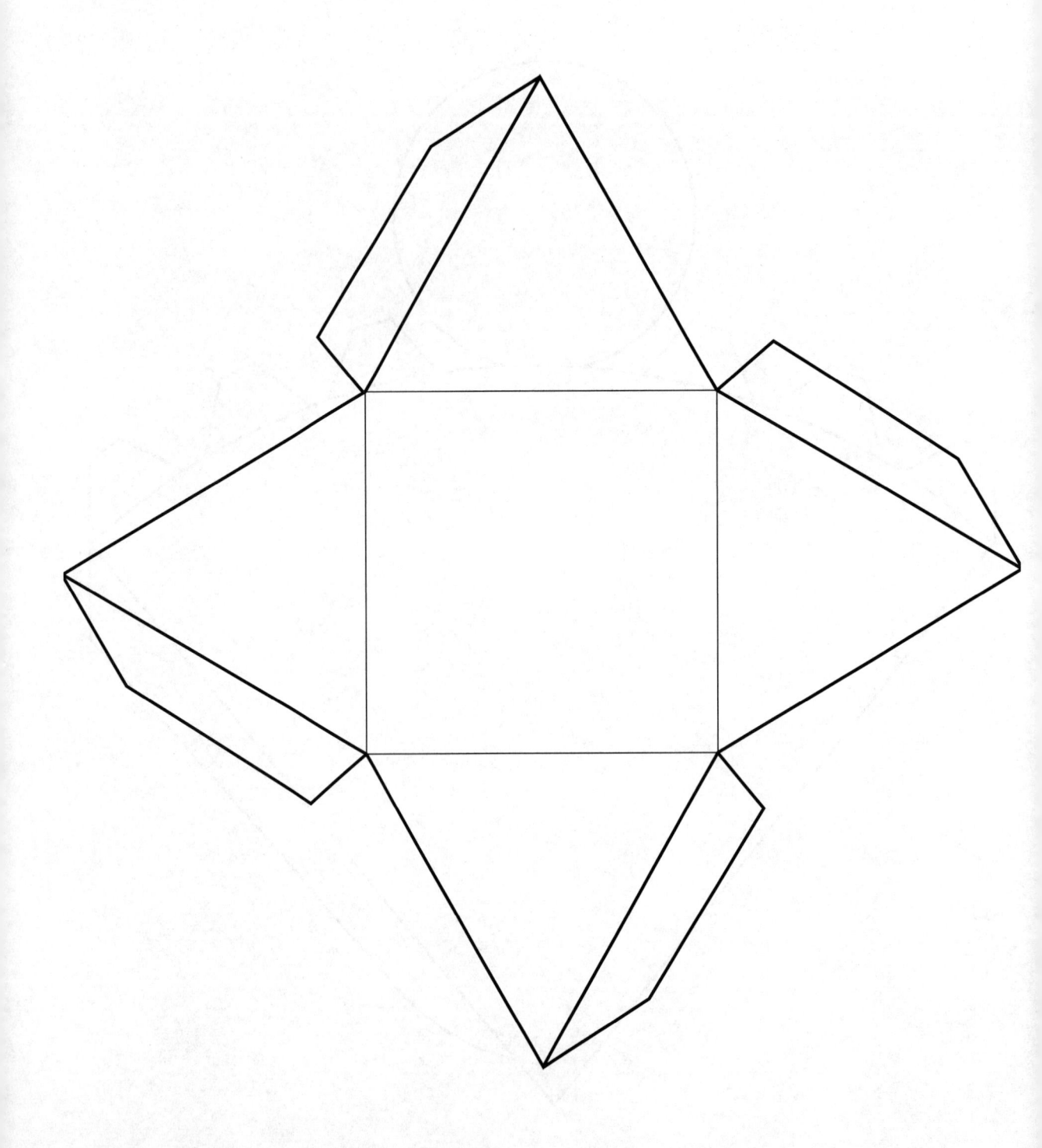

TR34 **Teacher's Resource Book**

Square Pyramid Pattern

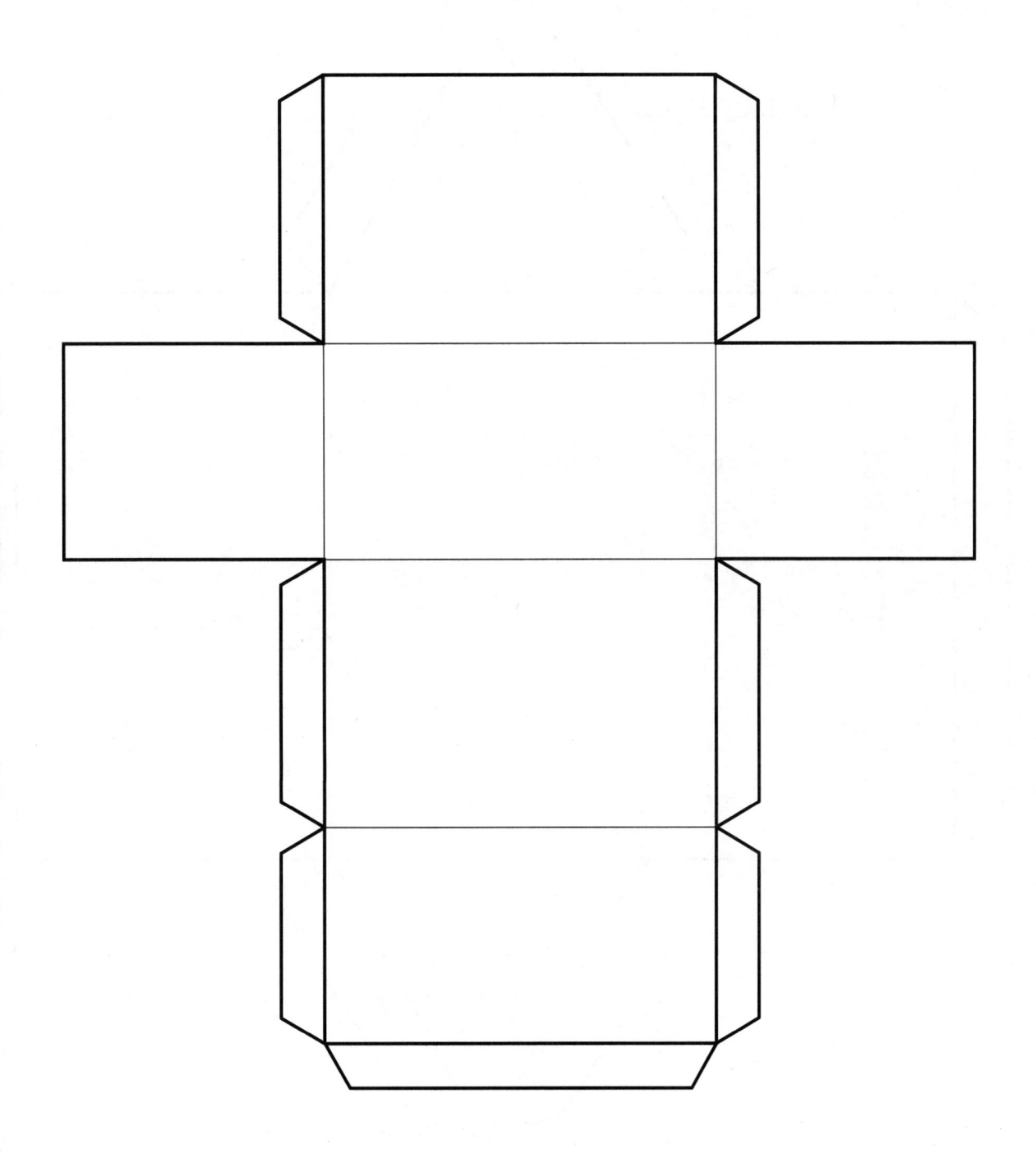

Rectangular Prism Pattern　　　　　　　　　　　　　　**Teacher's Resource Book　TR35**

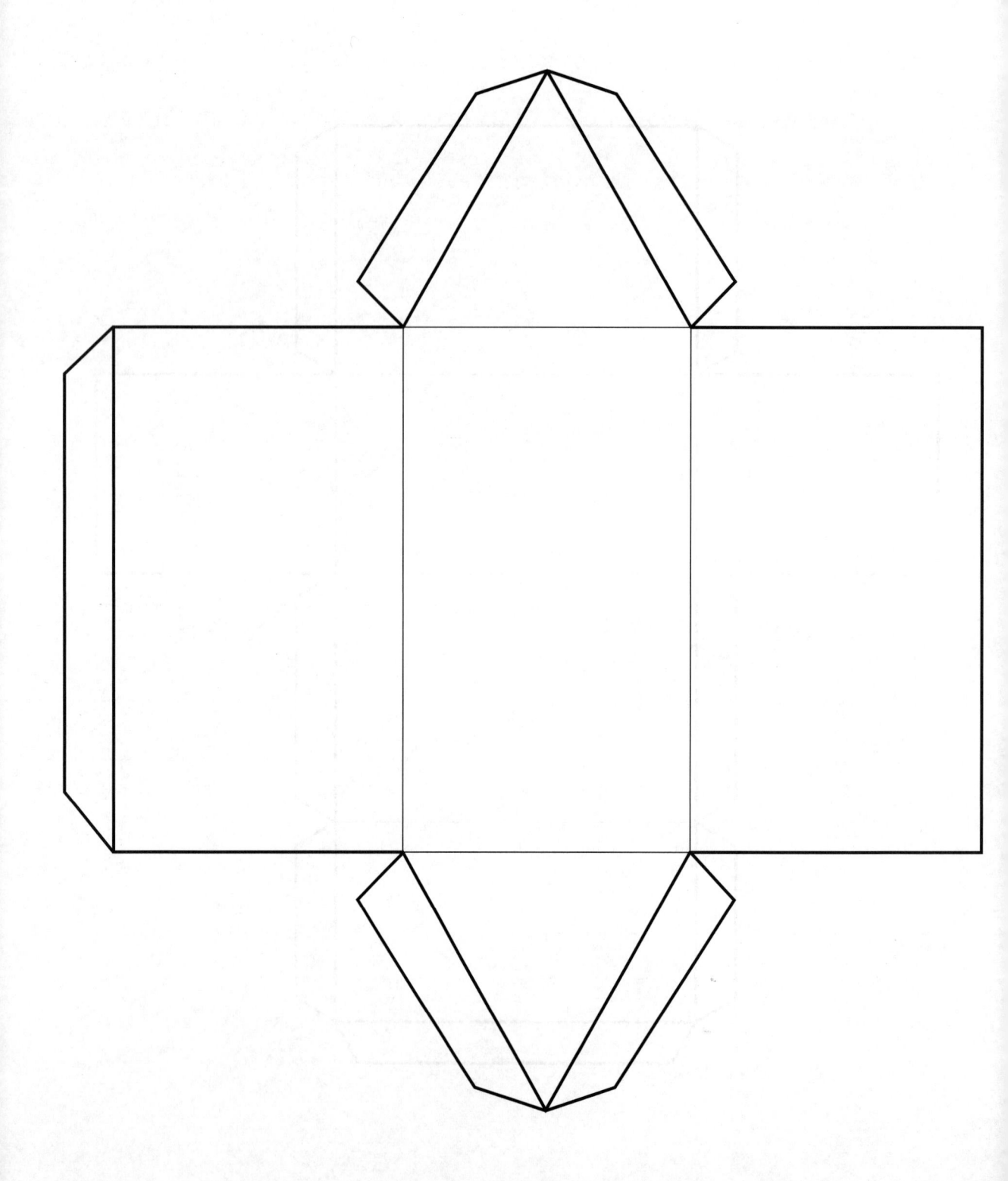

TR36 Teacher's Resource Book

Triangular Prism Pattern

Plane Figures

Teacher's Resource Book TR37

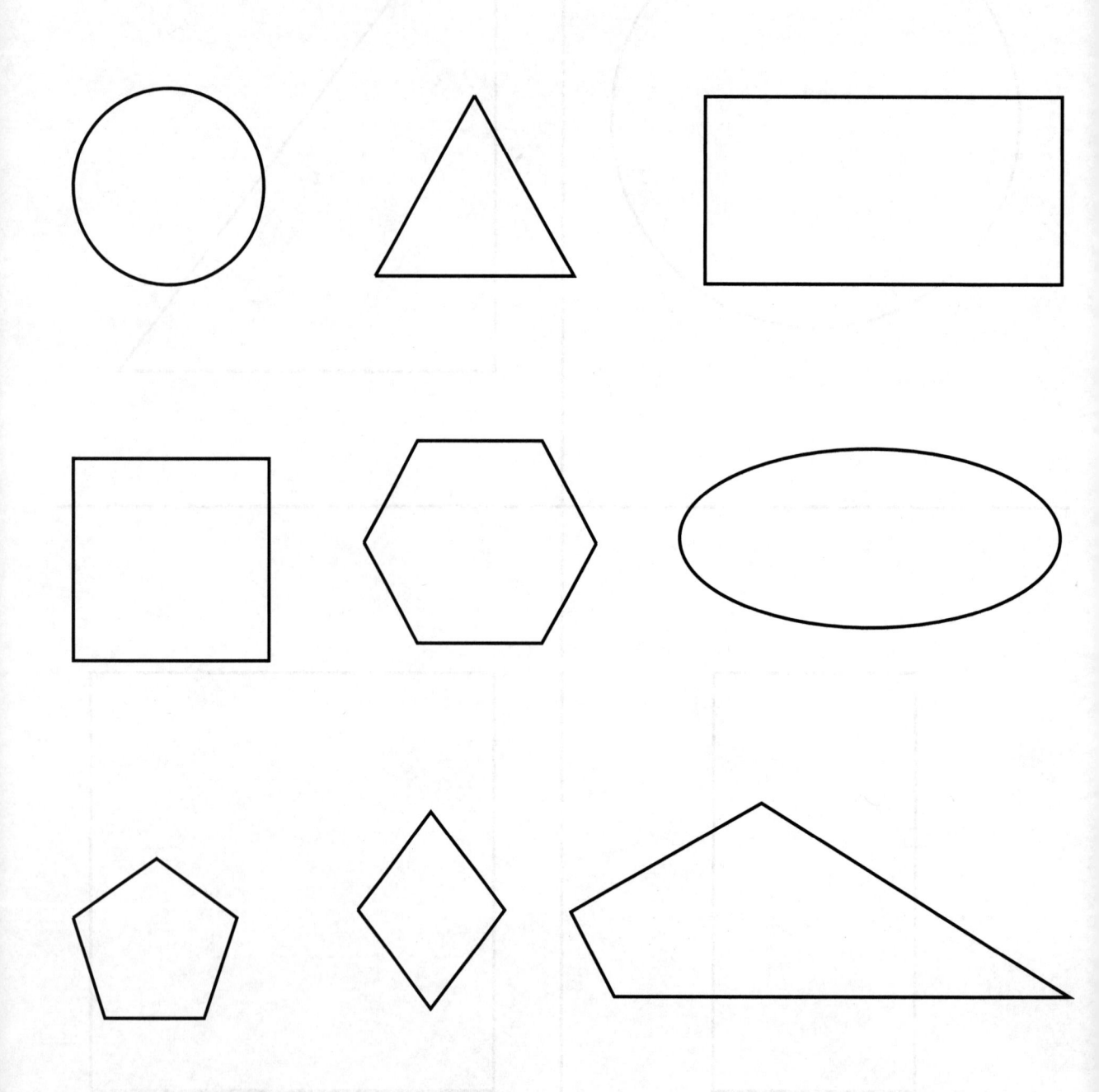

TR38 **Teacher's Resource Book**

Plane Figures

Circles　　　　　　　　　　　　　　　　　　　　　　　　**Teacher's Resource Book　TR39**

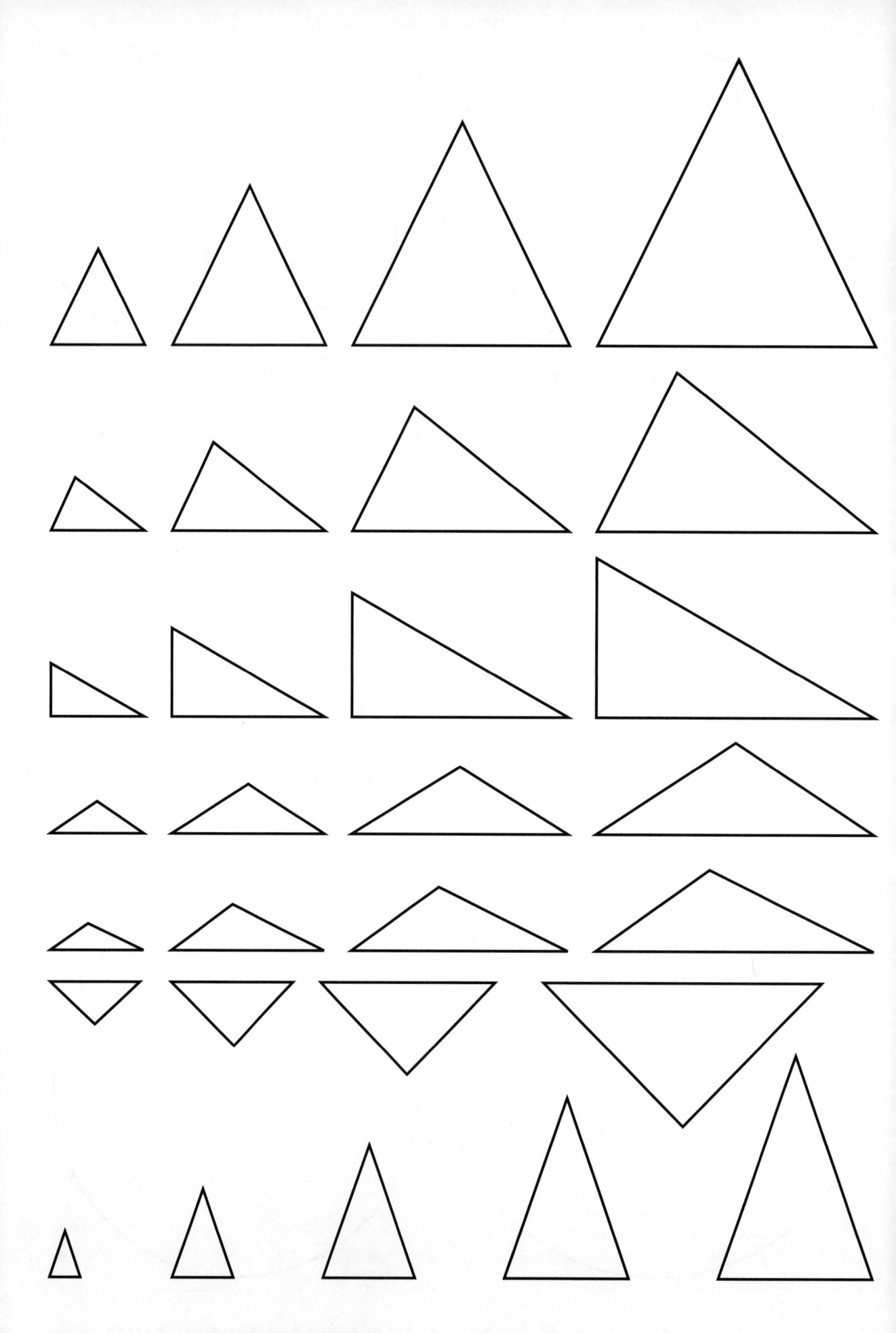

TR40 Teacher's Resource Book

Triangles

Squares and Rectangles   Teacher's Resource Book   **TR41**

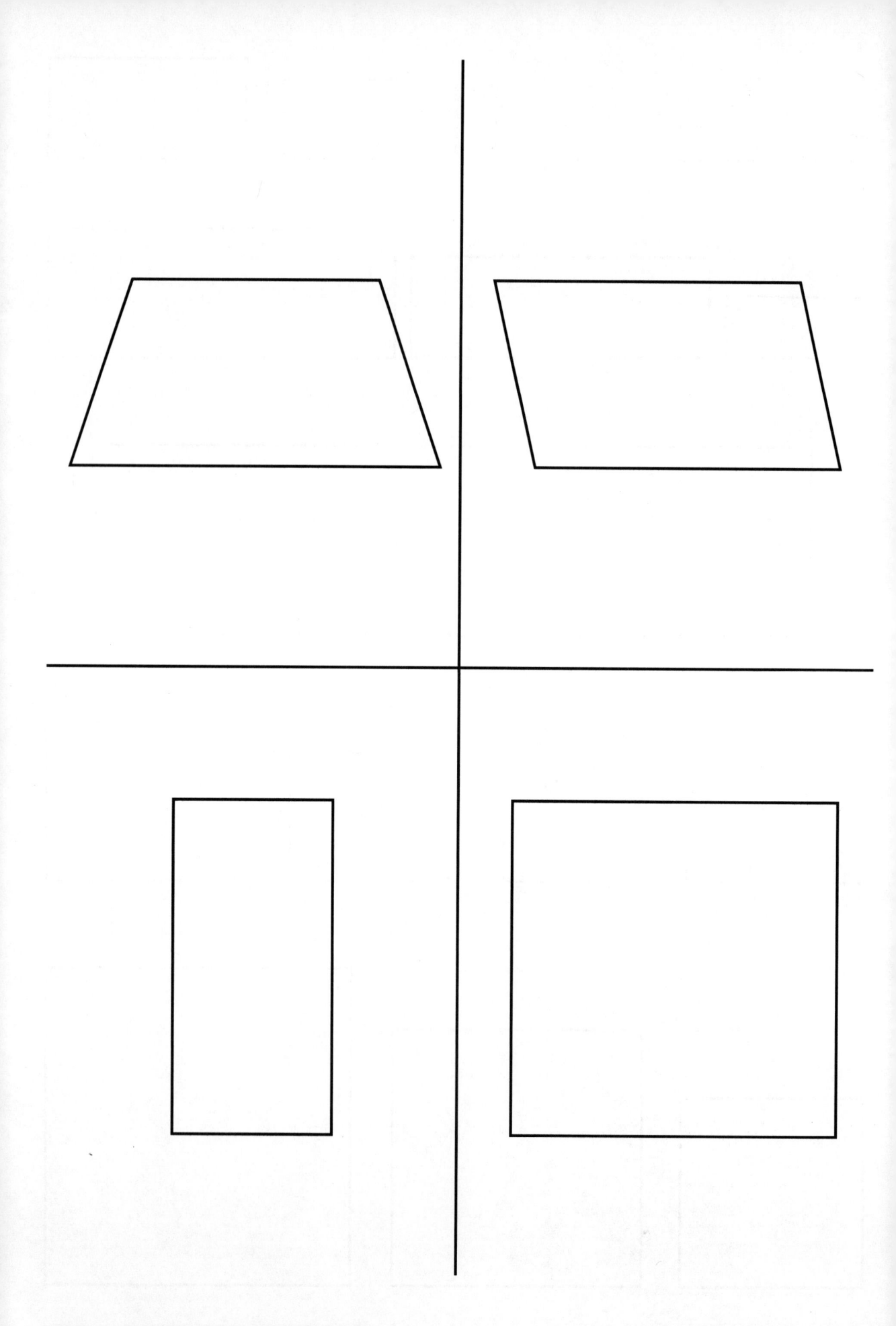

TR42 **Teacher's Resource Book**   Polygons: Quadrilaterals

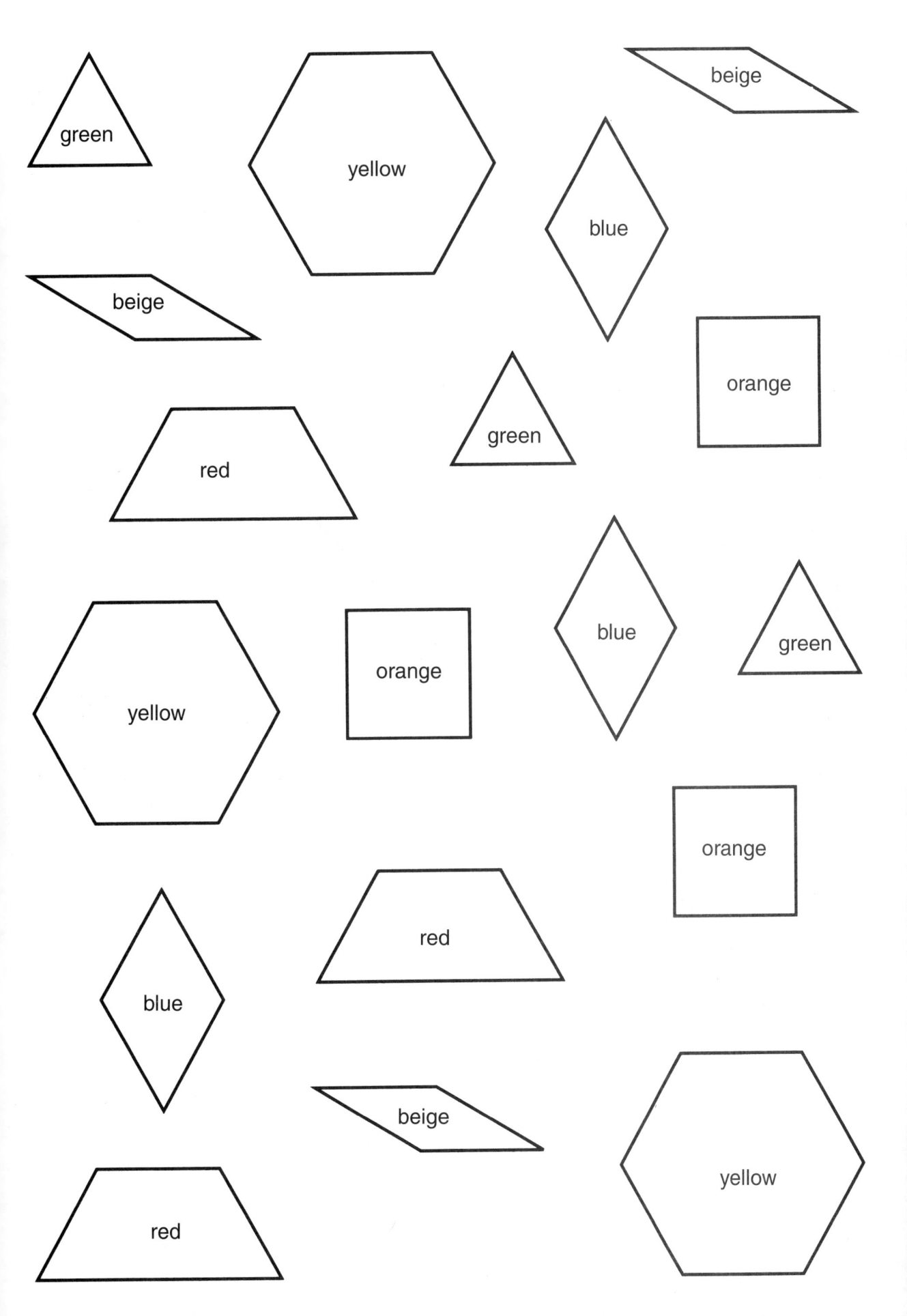

Pattern Block Patterns   Teacher's Resource Book   **TR 43**

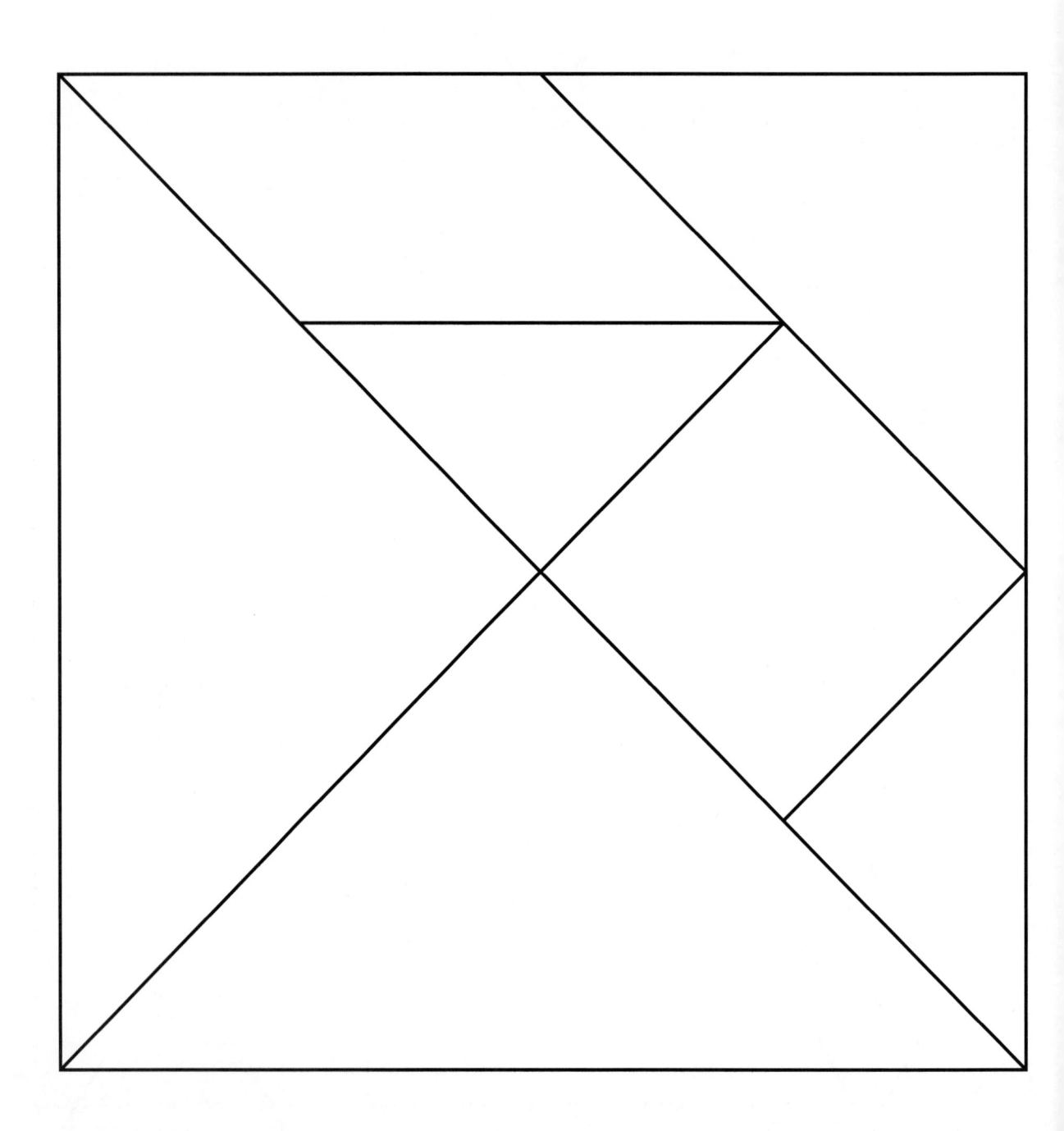

TR44   Teacher's Resource Book

Tangram Pattern

	Sunday	Monday	Tuesday	Wednesday	Thursday	Friday	Saturday

Blank Calendar

January

1	2	3	4	5	6	7
8	9	10	11	12	13	14
15	16	17	18	19	20	21
22	23	24	25	26	27	28
29	30	31				

February

			1	2	3	4
5	6	7	8	9	10	11
12	13	14	15	16	17	18
19	20	21	22	23	24	25
26	27	28				

March

			1	2	3	4
5	6	7	8	9	10	11
12	13	14	15	16	17	18
19	20	21	22	23	24	25
26	27	28	29	30	31	

April

						1
2	3	4	5	6	7	8
9	10	11	12	13	14	15
16	17	18	19	20	21	22
23	24	25	26	27	28	29
30						

May

	1	2	3	4	5	6
7	8	9	10	11	12	13
14	15	16	17	18	19	20
21	22	23	24	25	26	27
28	29	30	31			

June

				1	2	3
4	5	6	7	8	9	10
11	12	13	14	15	16	17
18	19	20	21	22	23	24
25	26	27	28	29	30	

Calendars: January–June

Large Analog Clock

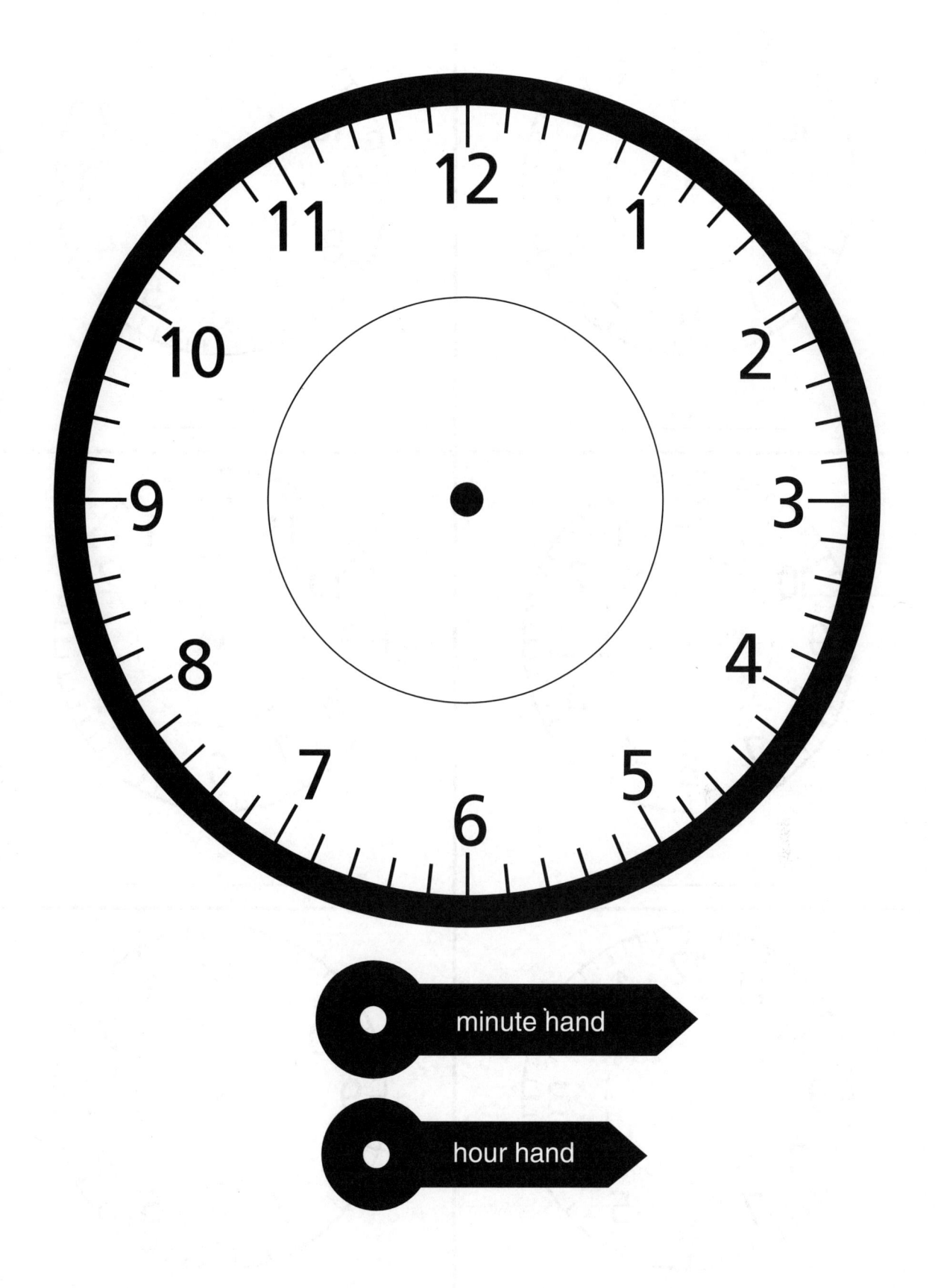

minute hand

hour hand

Teacher's Resource Book **TR47**

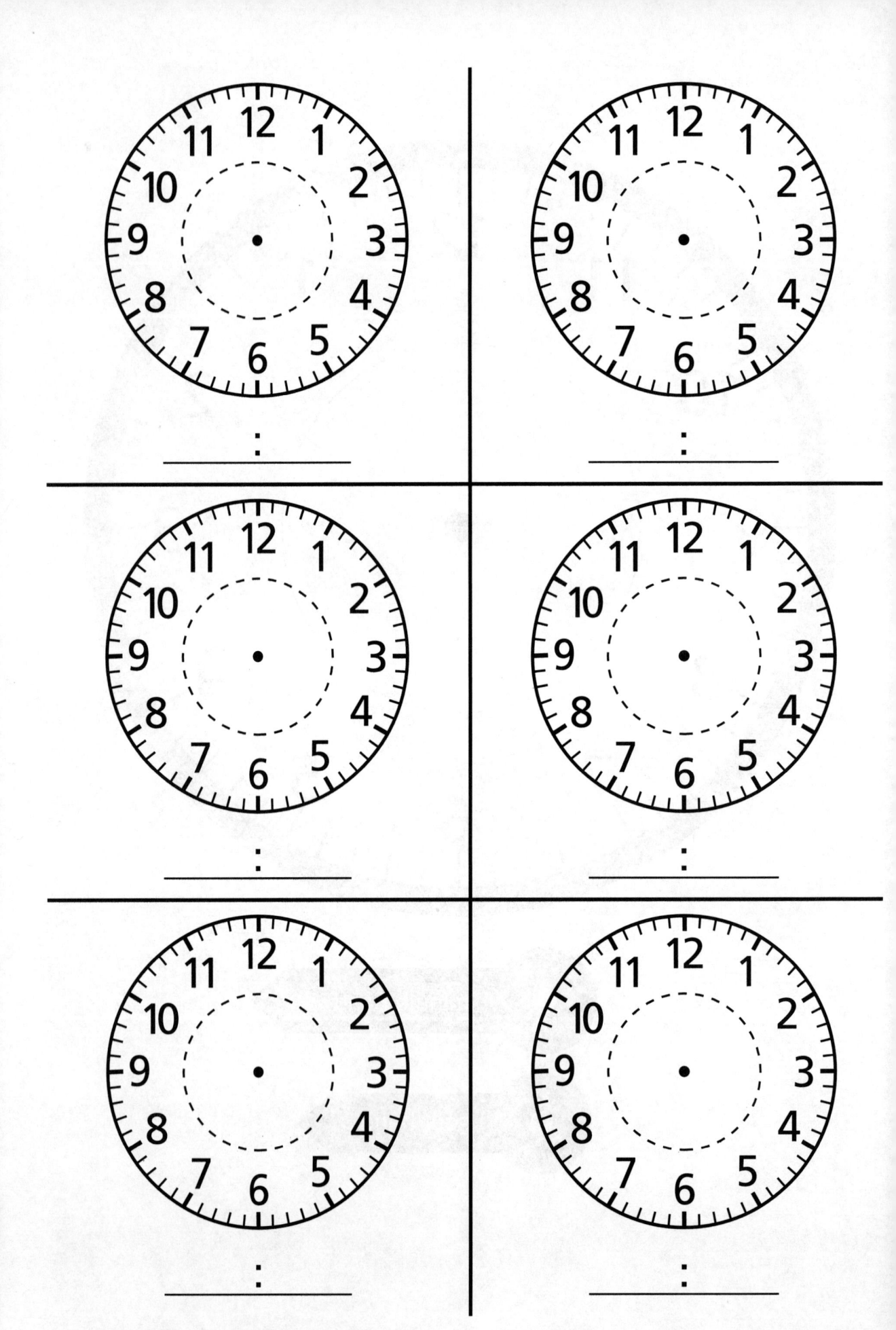

TR48 Teacher's Resource Book

Analog Clockfaces

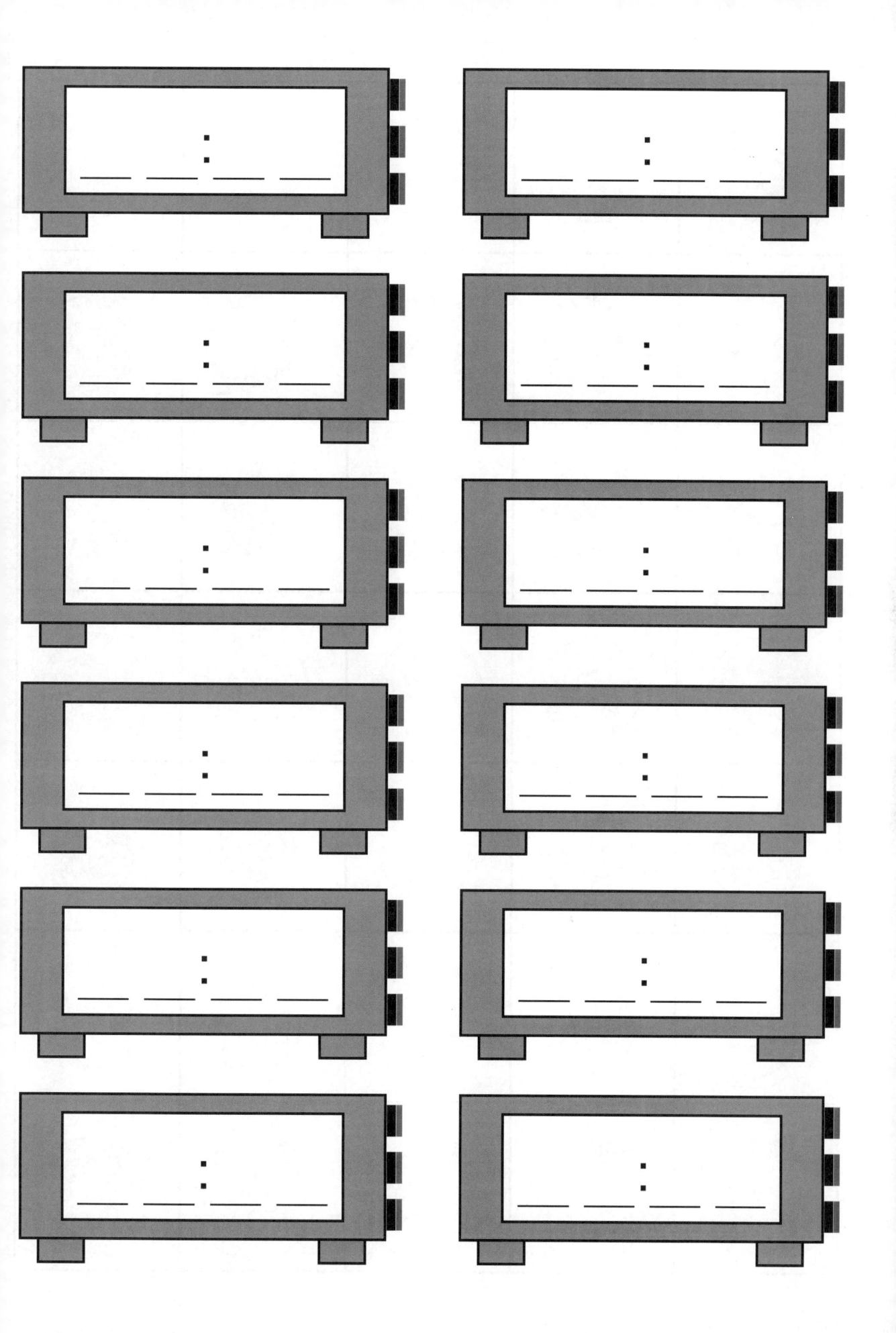

Digital Clock Model

Teacher's Resource Book TR49

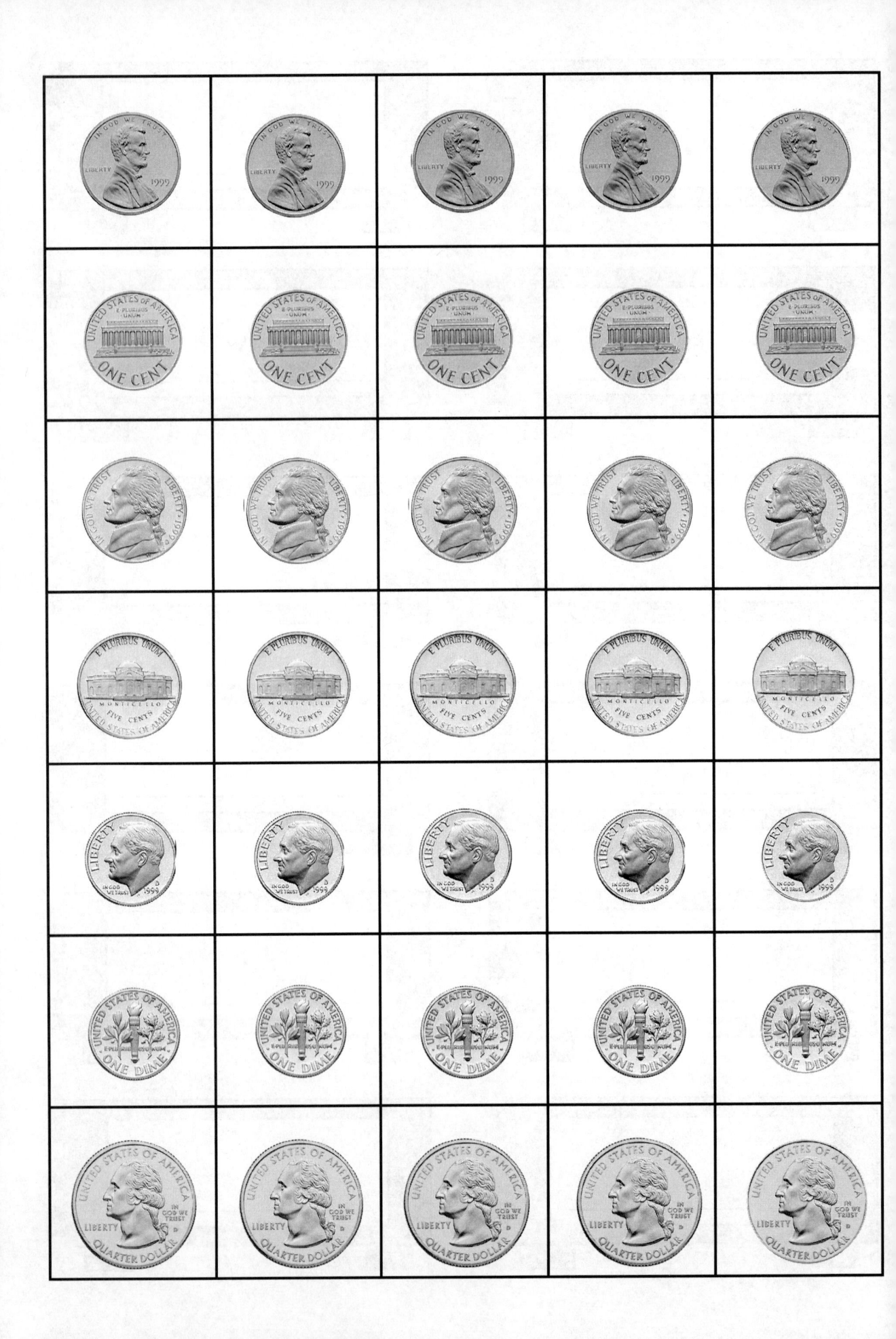

TR50 Teacher's Resource Book

Coins

Coins and Bills

Teacher's Resource Book TR51

TR52   Teacher's Resource Book

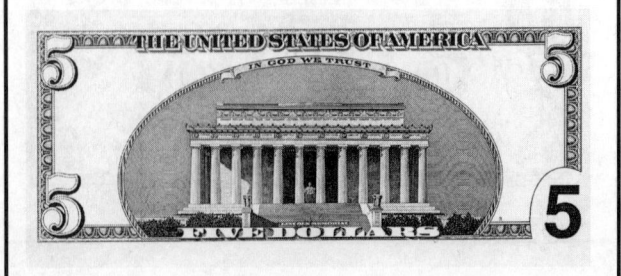

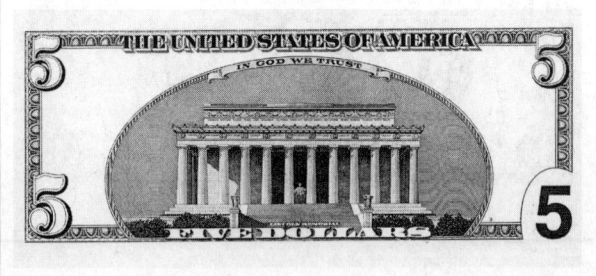

Bills

Workmat 4

Penny	
Nickel	
Dime	
Quarter	

Money Workmat 4 Teacher's Resource Book TR53

TR54 Teacher's Resource Book

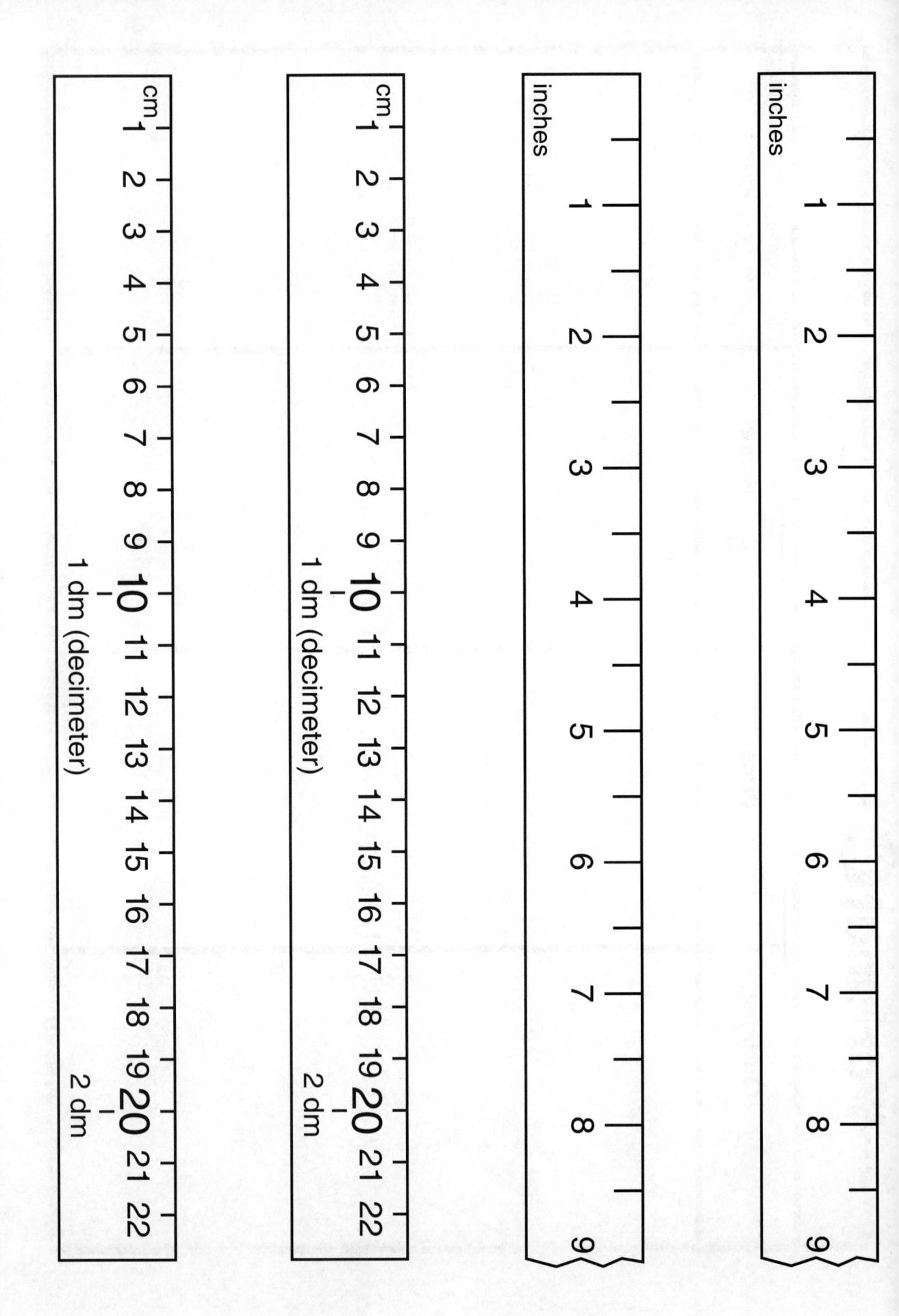

Rulers

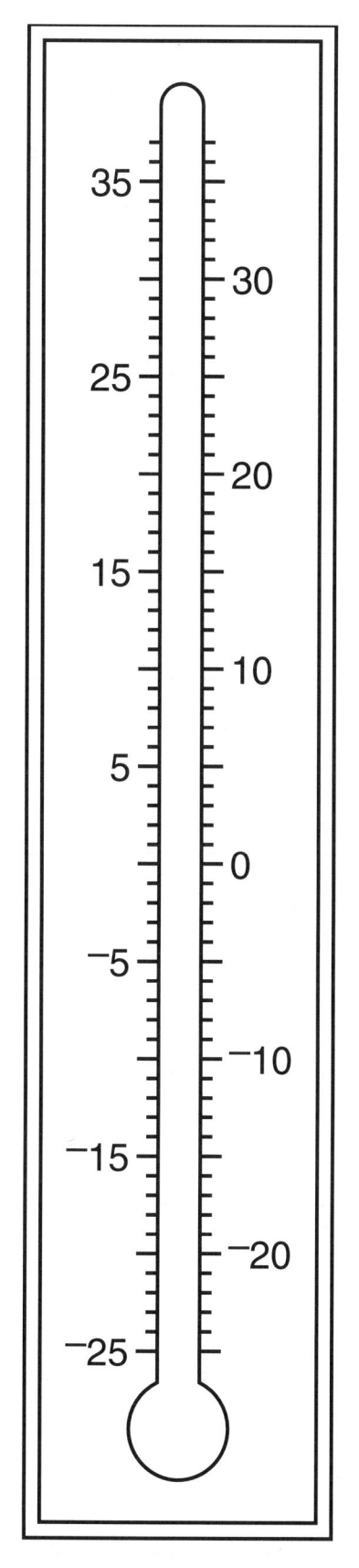

Celsius

____ °C

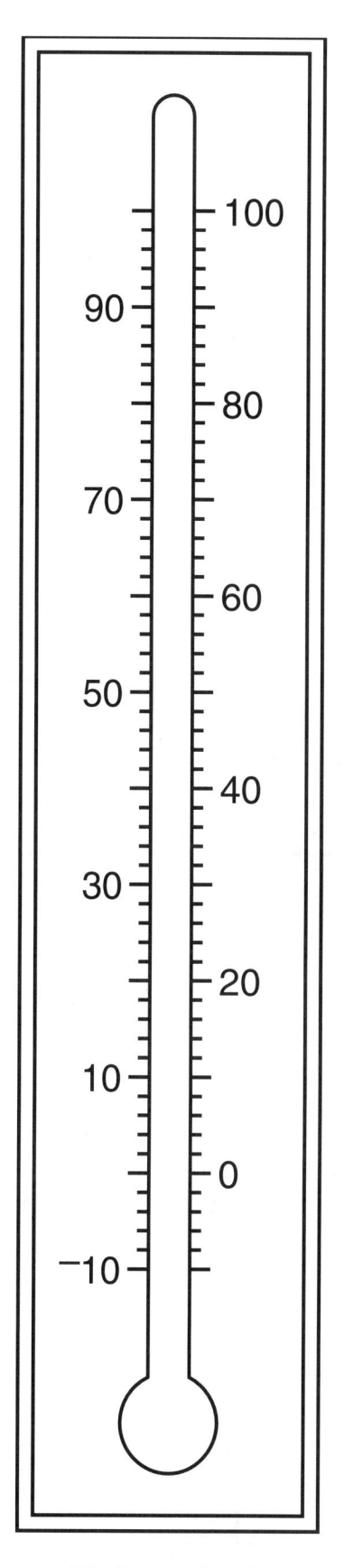

Fahrenheit

____ °F

Thermometers

Teacher's Resource Book TR55

TR56 **Teacher's Resource Book**　　　　　　　　　　1-Inch Grid Paper

1-Centimeter Grid Paper

Title: _____

TR58 Teacher's Resource Book

Pictograph Pattern

Title: _____

Scale Label: _____

Data Label: _____

Bar Graph Pattern 1 — Teacher's Resource Book **TR59**

Title_____

0

TR60　**Teacher's Resource Book**　　　　　　　　　　Bar Graph Pattern 2

Grid of Quadrant 1   Teacher's Resource Book   TR61

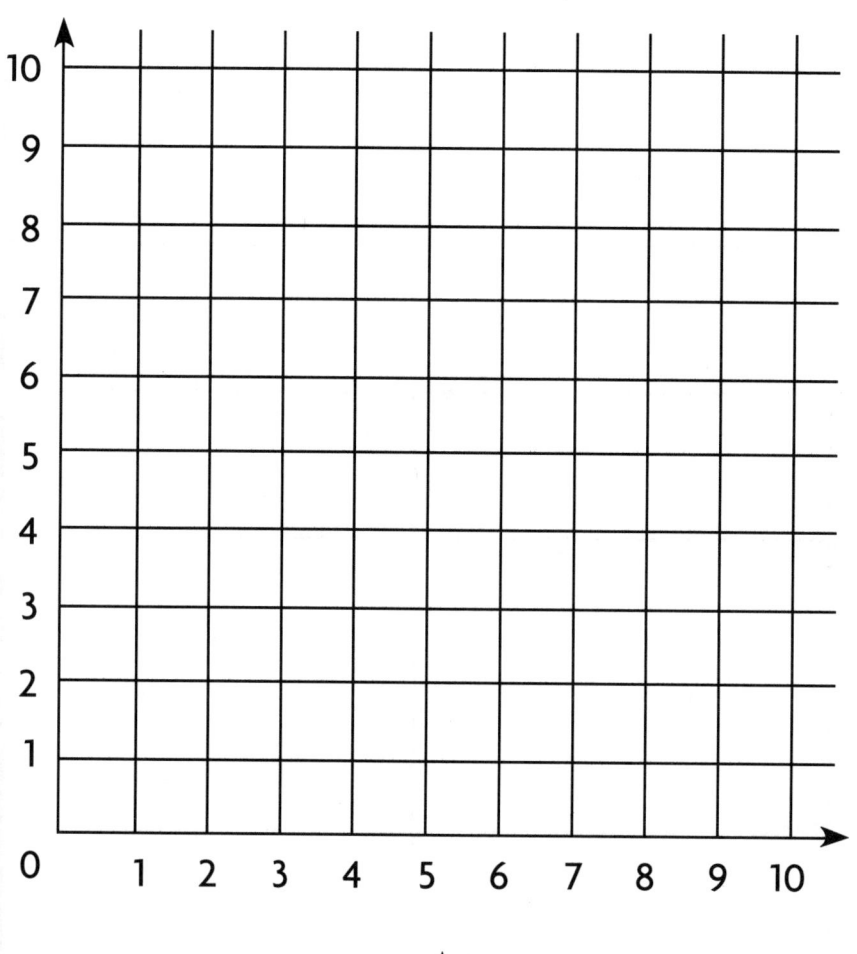

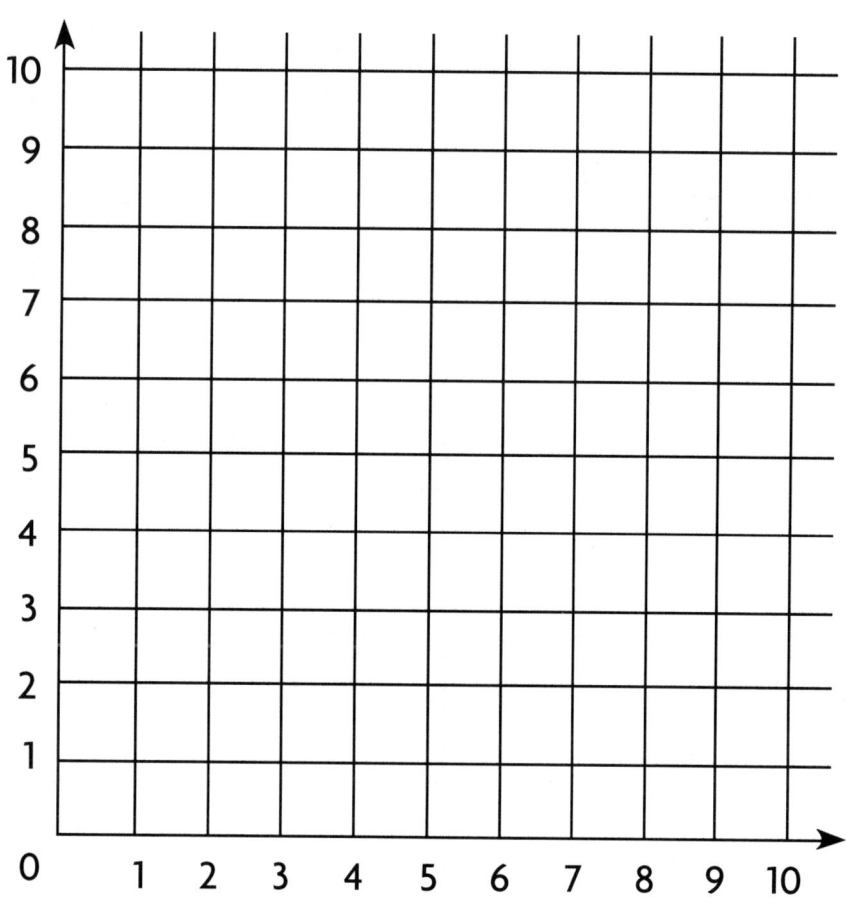

	Tally	Number

Tally Table

Title _____

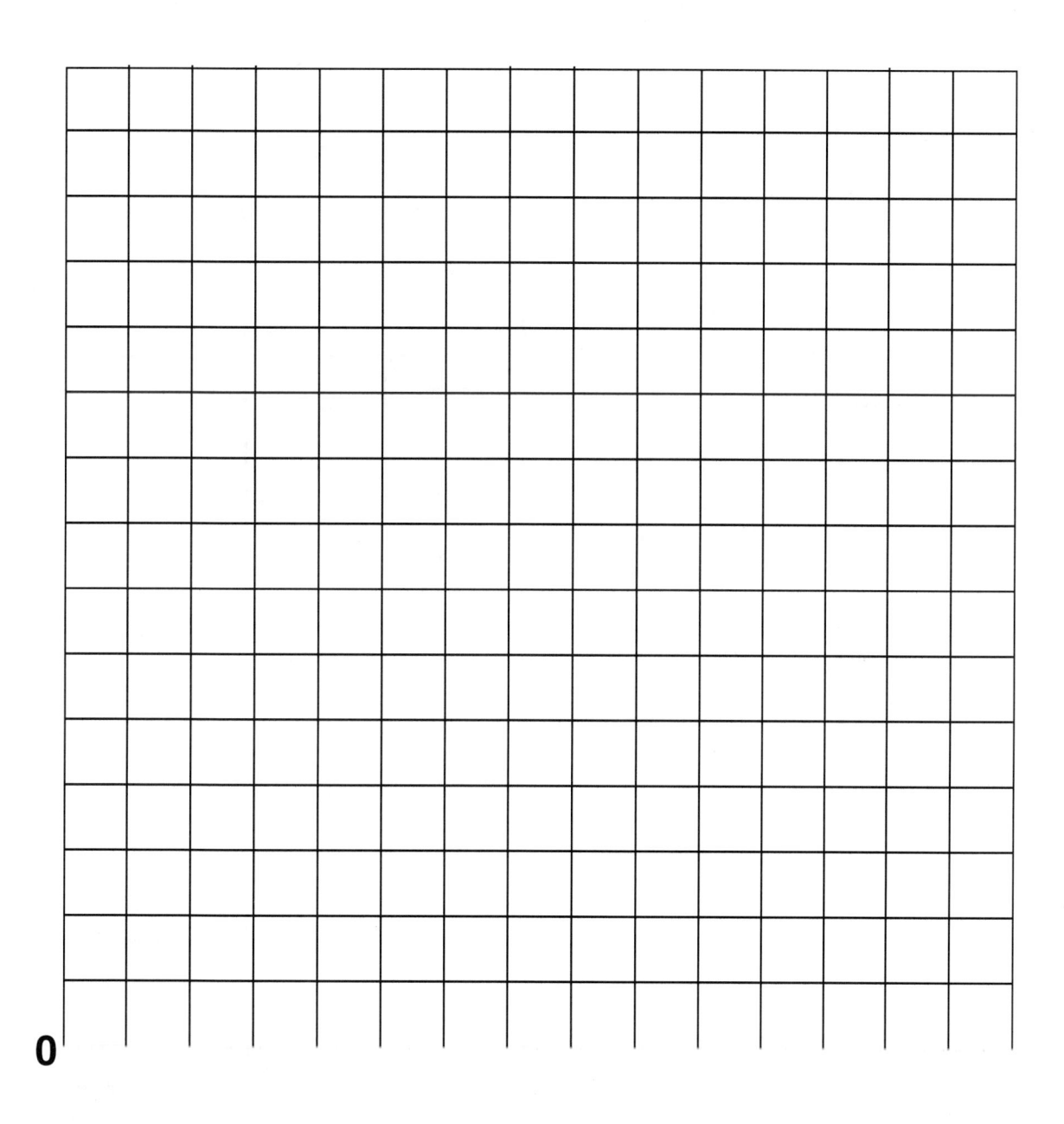

0

Line Graph Pattern

Teacher's Resource Book **TR63**

TR64 Teacher's Resource Book

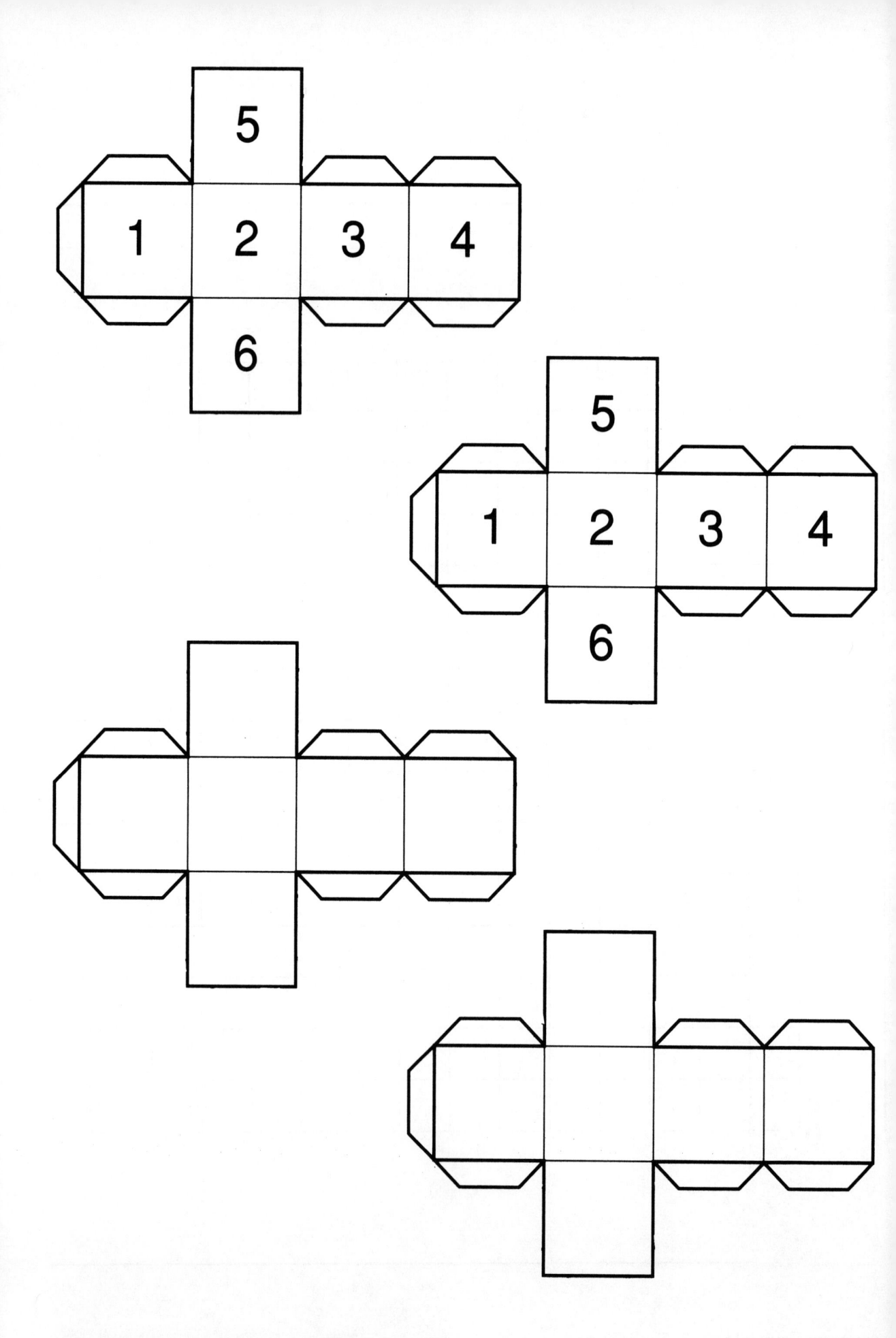

Number Cube Patterns

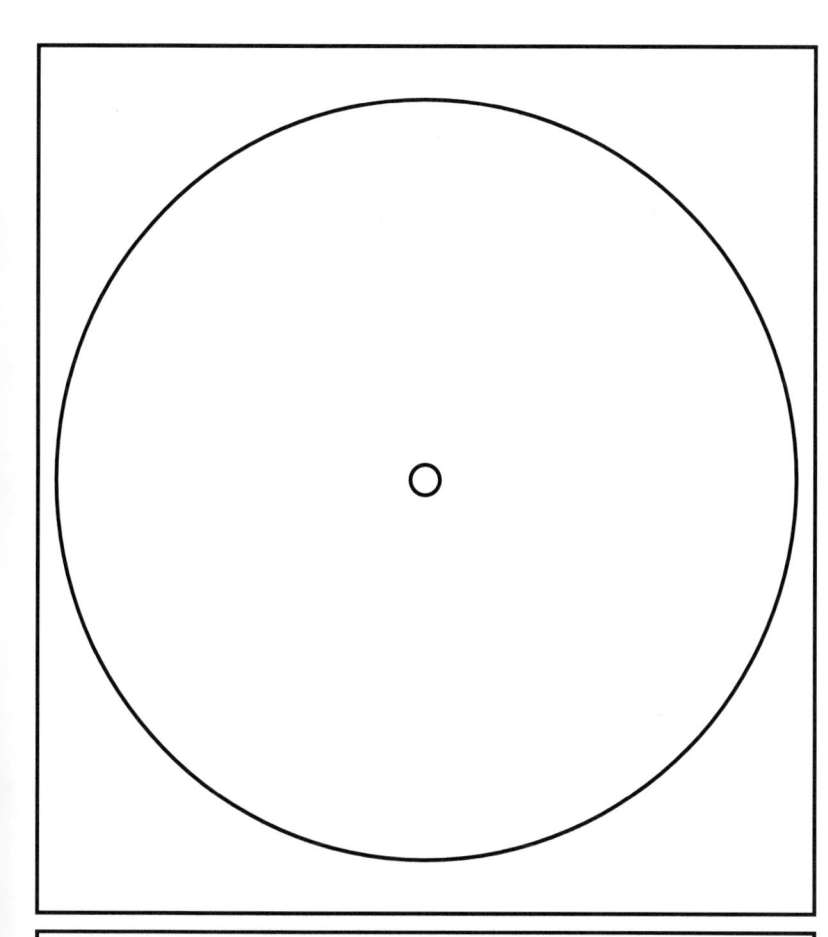

Spinner Tips

How to assemble spinner.
- Glue patterns to tagboard.
- Cut out and attach pointer with a fastener.

Alternative
- Students can use a paper clip and pencil instead.

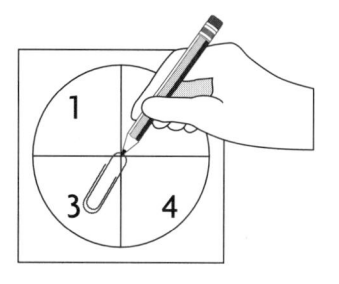

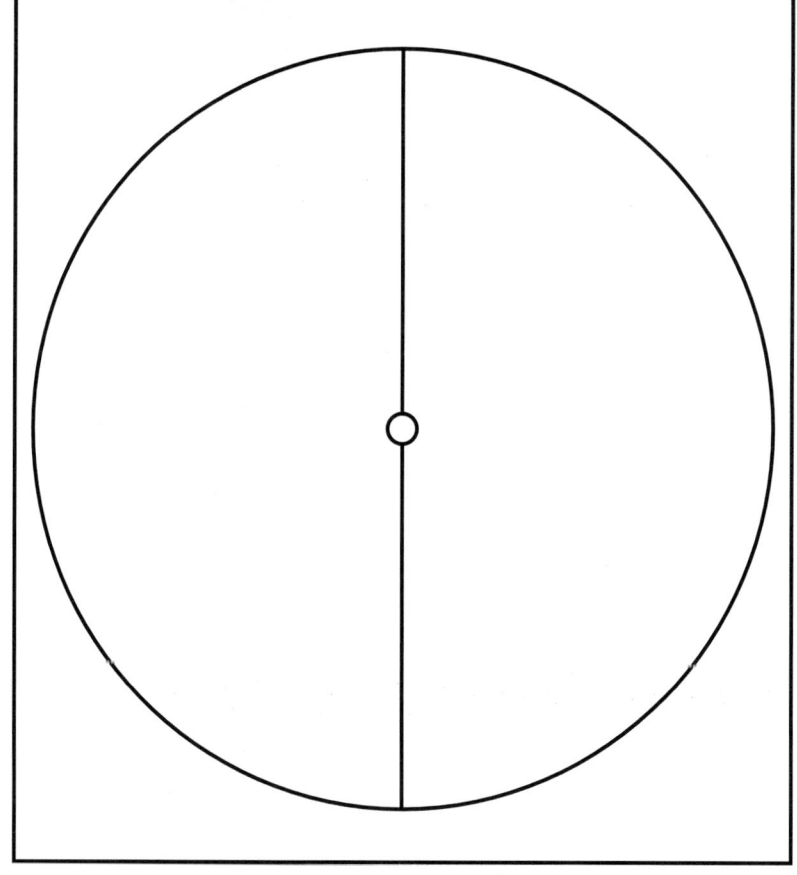

Spinners (blank and 2-section)

Teacher's Resource Book   TR65

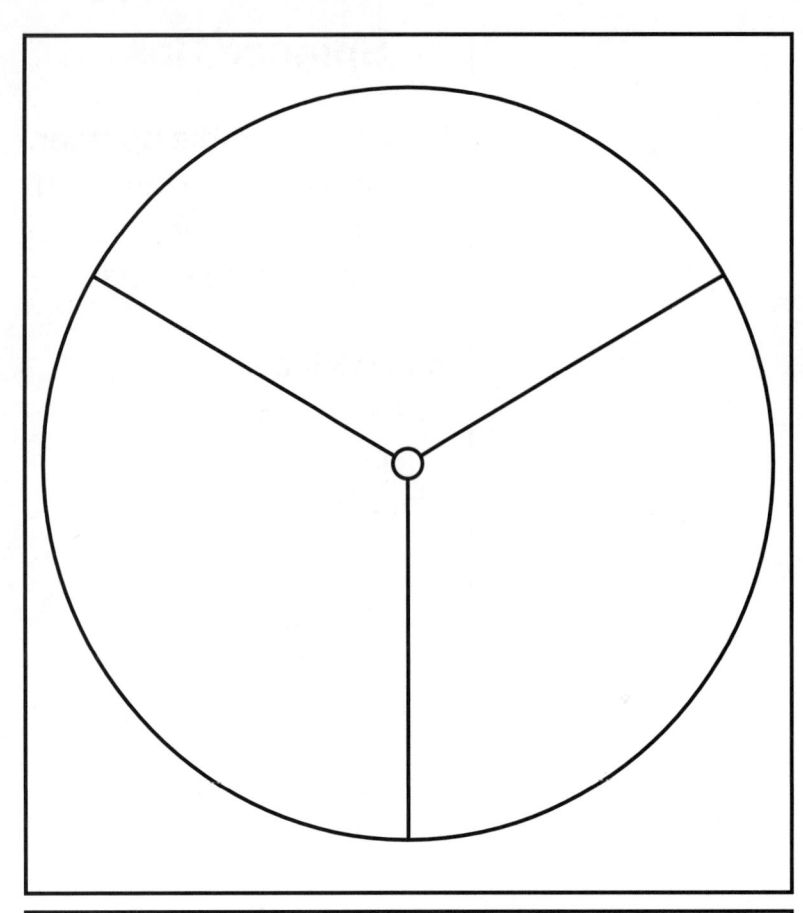

Spinner Tips

How to assemble spinner.
- Glue patterns to tagboard.
- Cut out and attach pointer with a fastener.

Alternative
- Students can use a paper clip and pencil instead.

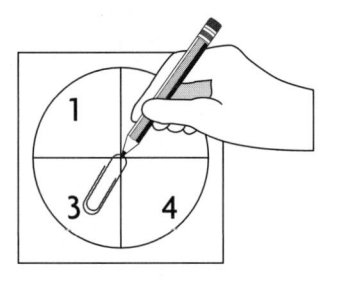

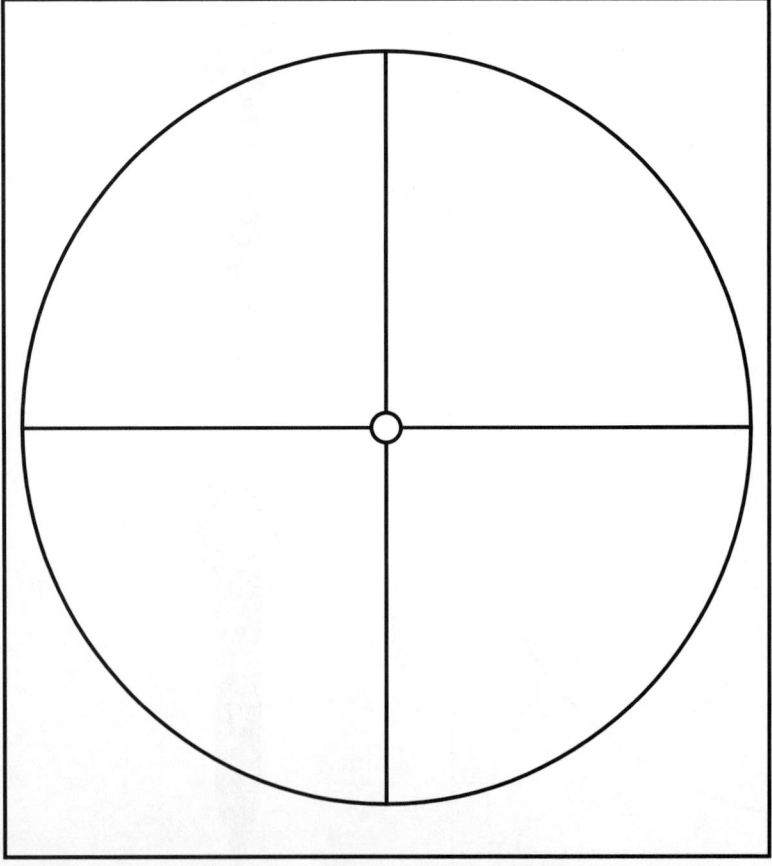

TR66   Teacher's Resource Book

Spinners (3- and 4-section)

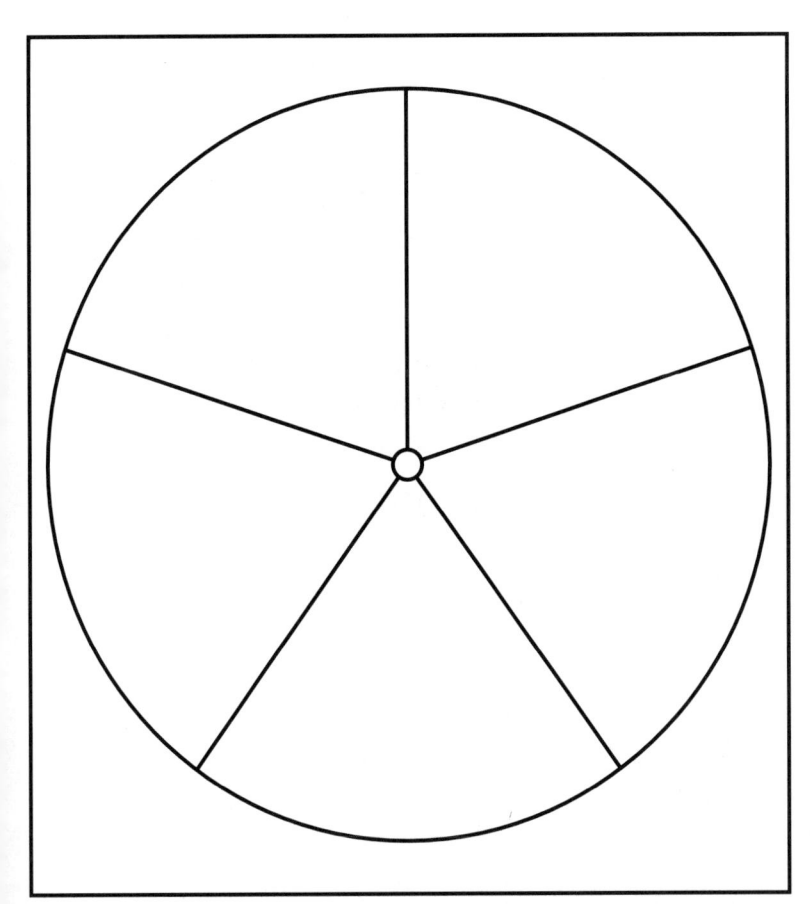

Spinner Tips

How to assemble spinner.
- Glue patterns to tagboard.
- Cut out and attach pointer with a fastener.

Alternative
- Students can use a paper clip and pencil instead.

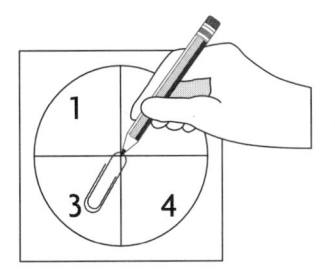

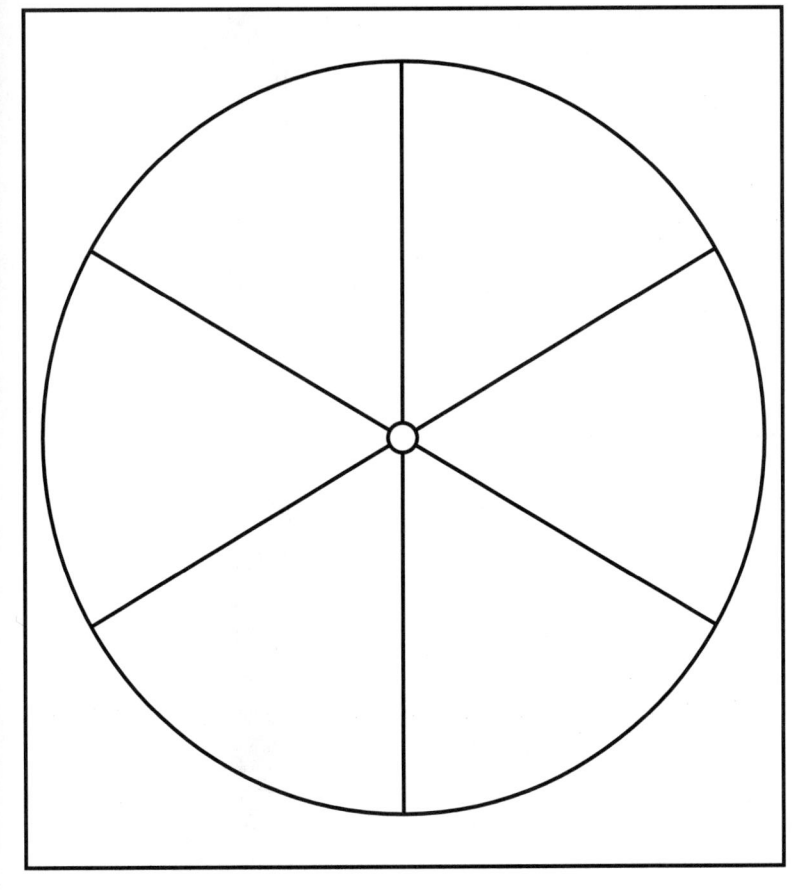

Spinners (5- and 6-section)

Teacher's Resource Book   TR67

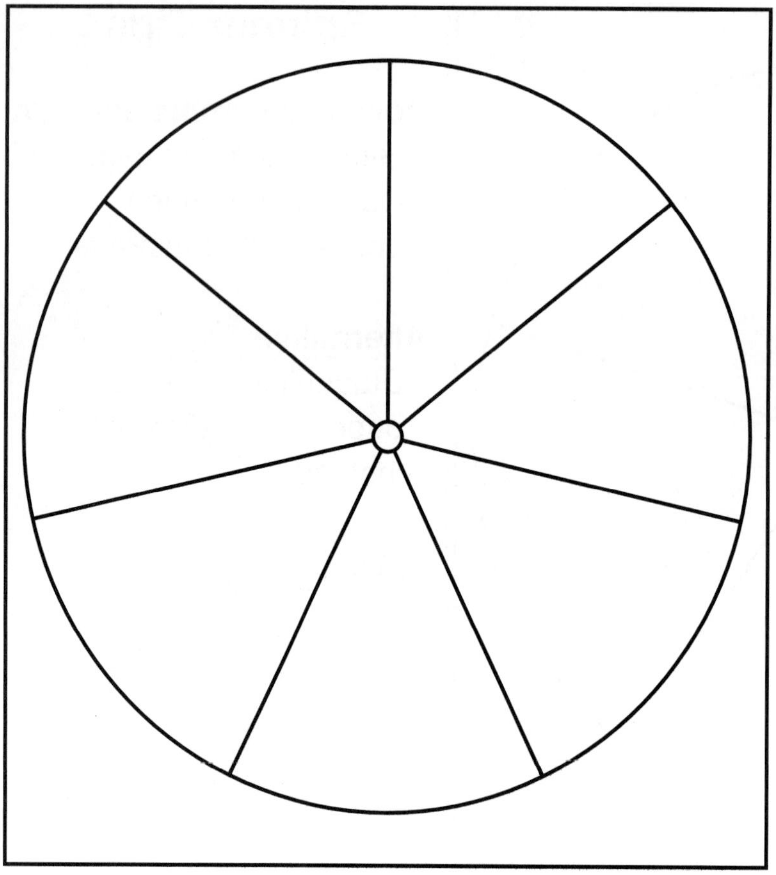

Spinner Tips

How to assemble spinner.
- Glue patterns to tagboard.
- Cut out and attach pointer with a fastener.

Alternative
- Students can use a paper clip and pencil instead.

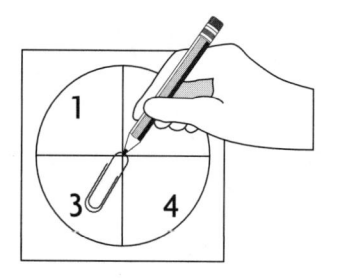

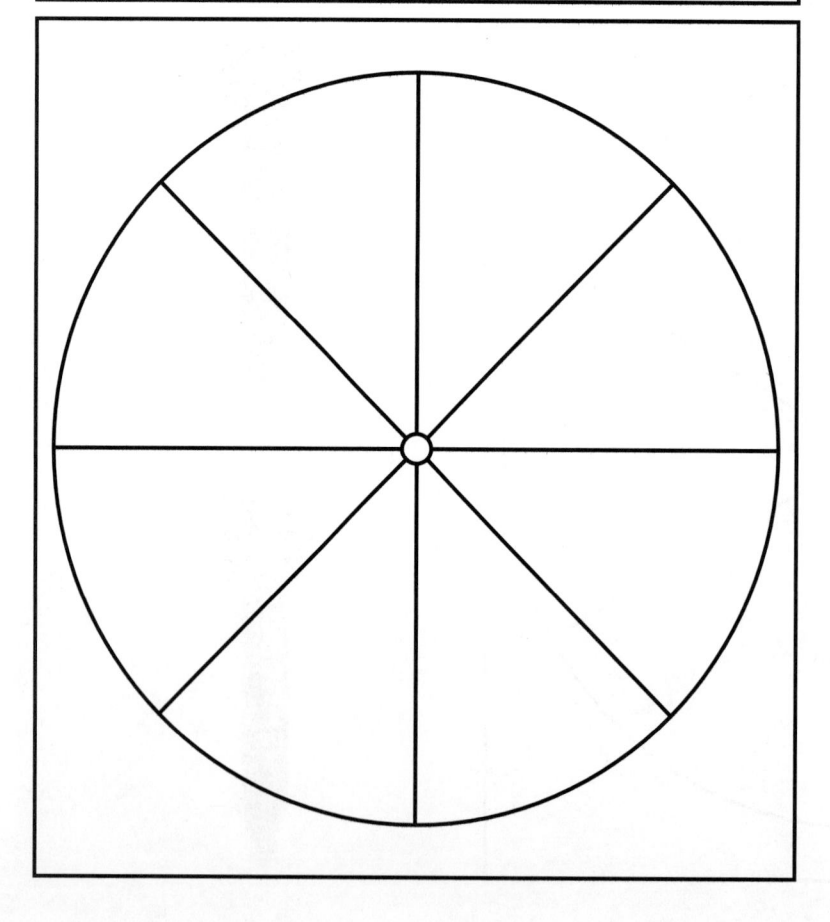

TR68 Teacher's Resource Book

Spinners (7- and 8-section)

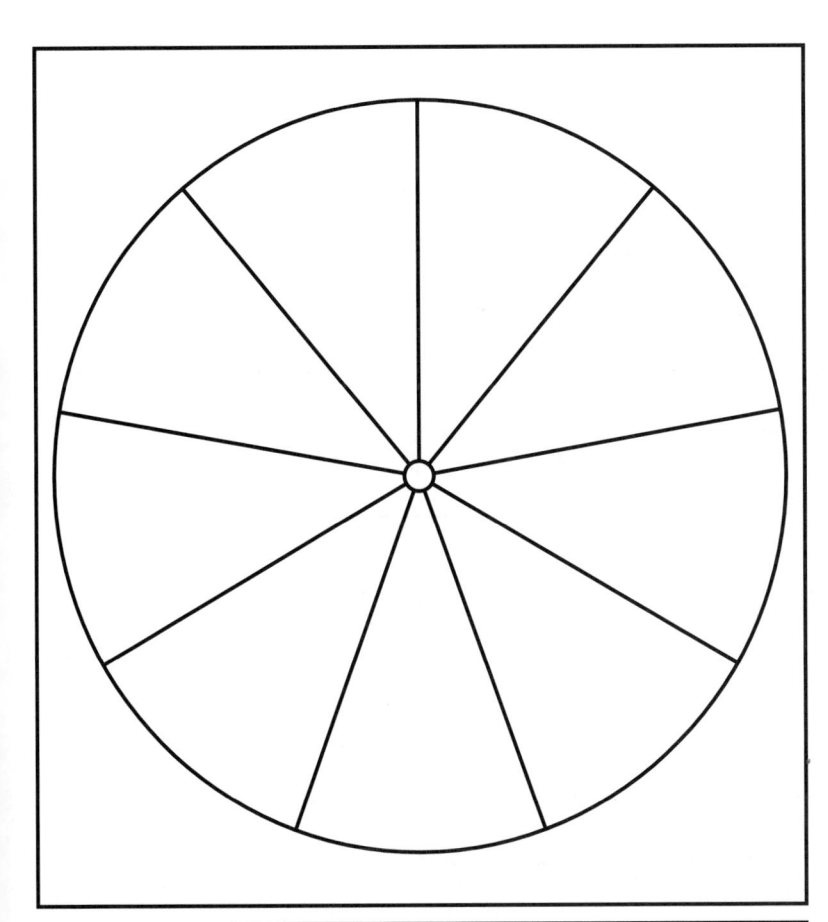

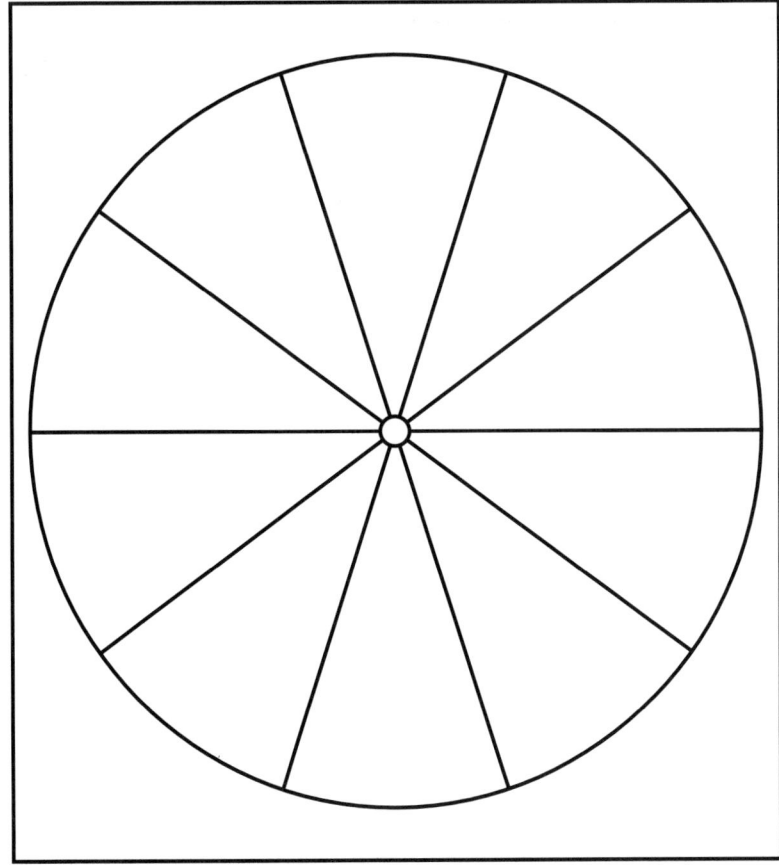

Spinner Tips

How to assemble spinner.
- Glue patterns to tagboard.
- Cut out and attach pointer with a fastener.

Alternative
- Students can use a paper clip and pencil instead.

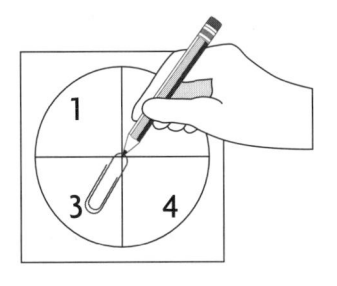

Spinners (9- and 10-section)

Teacher's Resource Book   TR69

High Number

Name _____

Round 1   ☐0,000 + ☐,000 + ☐00 + ☐0 + ☐ =

Round 2   ☐0,000 + ☐,000 + ☐00 + ☐0 + ☐ =

Round 3   ☐0,000 + ☐,000 + ☐00 + ☐0 + ☐ =

Round 4   ☐0,000 + ☐,000 + ☐00 + ☐0 + ☐ =

Round 5   ☐0,000 + ☐,000 + ☐00 + ☐0 + ☐ =

Activity Card 1

1. Say the twelve months of the year in order, three times.

Activity Card 2

2. Use the movie schedule to answer both questions.

Movie Schedule	Day	Time
The Big Cat	Fri	4:00–6:00
The Red Tomato	Sat	4:00–5:45
Lizard	Sat	5:30–7:15
My Old Shoe	Sun	4:00–5:30

Aunt Margaret will take you to the movies at 4:00 P.M. on Saturday. Which movie will you see? How long does the movie last?

Activity Card 3

3. Use the schedule to answer the questions.

Monday's Schedule	
Activity	Time
Morning announcements	8:15–8:30
Reading	8:30–9:15
Music	9:15–10:00
Language arts	10:00–11:00
Science	11:00–11:30
Lunch	11:30–12:15

How long is Science class? Which activities are 45 minutes long?

Activity Card 4

4. Use the calendar to answer the questions.

January						
Sun	Mon	Tue	Wed	Thur	Fri	Sat
			1	2	3	4
5	6	7	8	9	10	11
12	13	14	15	16	17	18
19	20	21	22	23	24	25
26	27	28	29	30	31	

On what day of the week is February 1? On what day of the week is New Year's Day? How many Saturdays are there in January?

Answer Card

1. January, February, March, April, May, June, July, August, September, October, November, December

2. *The Red Tomato*; 1 hour 45 minutes

3. 30 minutes; reading, music, lunch

4. Feb. 1 is Saturday; New Year's is Wednesday; 4 Saturdays

Beat the Clock

Around the Moon

0	1	2	3	4
5	<u>6</u>	7	8	<u>9</u>

TR72 Teacher's Resource Book

Around the Moon

45 ÷ 5	45 ÷ 9	9 × 5	5 × 9
35 ÷ 5	35 ÷ 7	7 × 5	5 × 7
28 ÷ 4	28 ÷ 7	7 × 4	4 × 7
12 ÷ 4	12 ÷ 3	3 × 4	4 × 3
27 ÷ 3	27 ÷ 9	3 × 9	9 × 3
20 ÷ 4	20 ÷ 5	4 × 5	5 × 4
18 ÷ 3	18 ÷ 6	6 × 3	3 × 6
6 ÷ 2	6 ÷ 3	2 × 3	3 × 2
10 ÷ 2	10 ÷ 5	5 × 2	2 × 5
8 ÷ 4	8 ÷ 2	4 × 2	2 × 4

Fact Family

Outcome Table

Spinner 1 | Spinner 2 | Spinner 3

Color	Prediction	Outcome	Color	Prediction	Outcome	Color	Prediction	Outcome

Spinner 1 | Spinner 2 | Spinner 3

Color	Prediction	Outcome	Color	Prediction	Outcome	Color	Prediction	Outcome

Spinner 1 | Spinner 2 | Spinner 3

Color	Prediction	Outcome	Color	Prediction	Outcome	Color	Prediction	Outcome

Angles from A to . . .

Start

| A | R | K | 5 | N | 8 | P | 9 |

5
3
M
6
W
2
X
Y
1

| F | C | 4 | 7 | E | O | H | Z |

Right Angle Ramble · Teacher's Resource Book · TR75

How would you measure each item?
Use centimeter (cm), decimeter (dm), or meter (m).

☐ 1. Measure length of swimming pool.

☐ 2. Measure length of referee's whistle.

☐ 3. Measure length of track for running a race.

☐ 4. Measure length of tennis racket.

☐ 5. Measure width of baseball diamond.

☐ 6. Measure width of bicycle tire.

☐ 7. Measure length of arrows for archery.

☐ 8. Measure size of table tennis balls.

☐ 9. Measure length of soccer field.

☐ 10. Measure width of wheels on roller skates.

☐ 11. Measure height of basketball goal.

☐ 12. Measure length of a paper clip.

....

Answer Key:
1. m
2. cm
3. m
4. dm
5. m
6. cm or dm
7. dm
8. cm
9. m
10. cm
11. m
12. cm

TR 76 Teacher's Resource Book   Ready, Set, Measure!

0.44	0.21
1.81	2.25
2.1	3.15
0.19	0.9
1.36	1.55
2.6	2.66
3.78	3.95
0.84	0.62
1.23	2.34
3.45	2.99

Color It In   **Teacher's Resource Book**

Number	Division Toss Divide by	Points
	2	
	3	
	4	
	Lucky Space Points = Number Tossed	
	5	
	6	
	7	
	Lucky Space Points = Double the Number Tossed	
	8	
	9	
	Total	

Remainders:

Number	Division Toss Divide by	Points
	2	
	3	
	4	
	Lucky Space Points = Number Tossed	
	5	
	6	
	7	
	Lucky Space Points = Double the Number Tossed	
	8	
	9	
	Total	

Remainders:

Number	Division Toss Divide by	Points
	2	
	3	
	4	
	Lucky Space Points = Number Tossed	
	5	
	6	
	7	
	Lucky Space Points = Double the Number Tossed	
	8	
	9	
	Total	

Remainders:

Number	Division Toss Divide by	Points
	2	
	3	
	4	
	Lucky Space Points = Number Tossed	
	5	
	6	
	7	
	Lucky Space Points = Double the Number Tossed	
	8	
	9	
	Total	

Remainders:

BINGO

		FREE		

Bingo Grid

Math Award

To _____

For _____

_____

_____

By _____

On this day of _____

EXCELLENCE

TR80 Teacher's Resource Book

Award Certificate

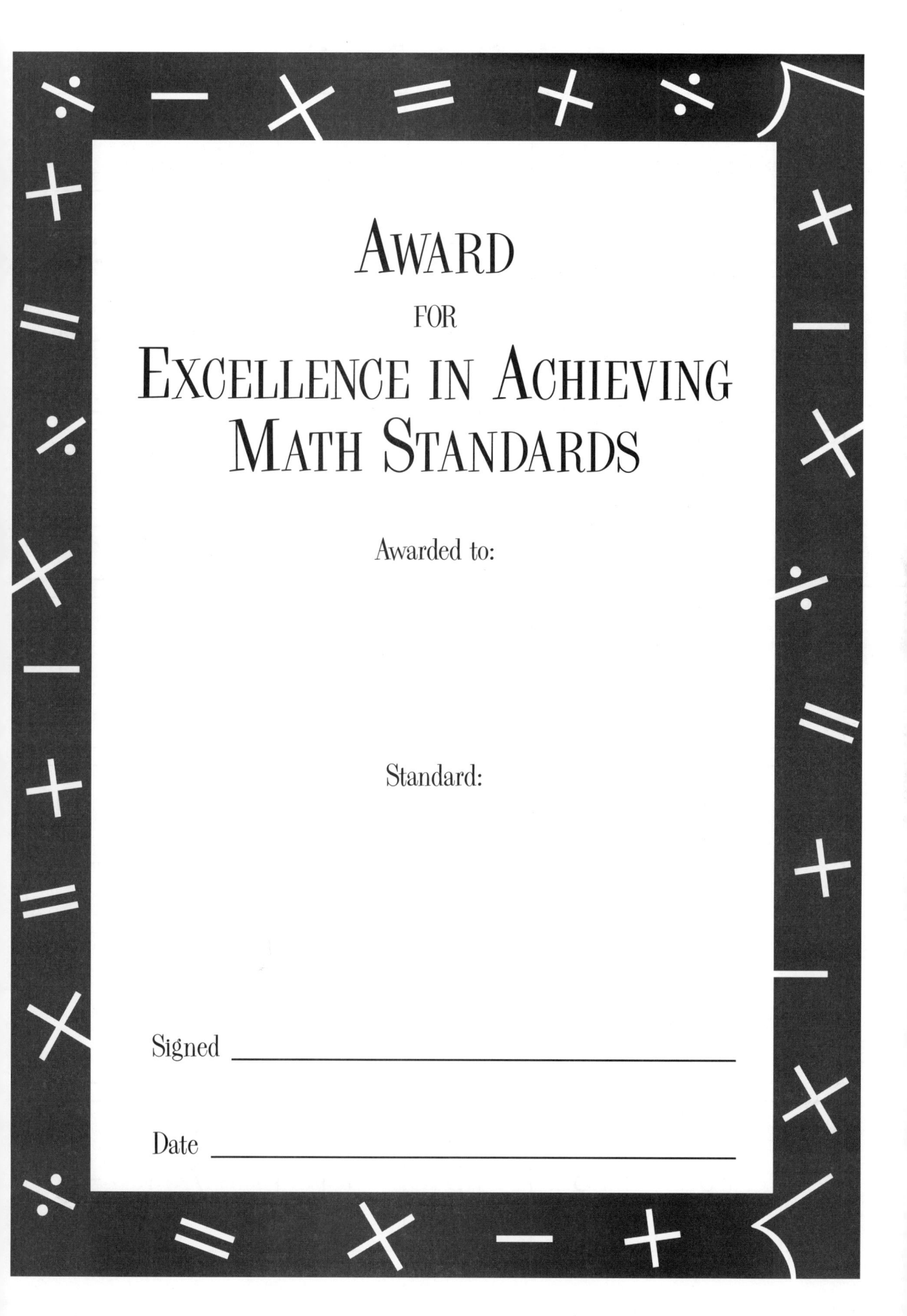

Award
for
Excellence in Achieving Math Standards

Awarded to:

Standard:

Signed _____

Date _____

Award Certificate Teacher's Resource Book **TR81**

Daily Facts Practice

A	1 +0	1 +6	2 +2	3 +2	0 +9	4 +1
B	5 +1	0 +0	1 +2	3 +3	4 +2	2 +4
C	0 +3	1 +1	2 +3	7 +0	2 +6	9 +2
D	2 −1	0 −0	7 −1	8 −0	2 −2	8 −2
E	4 −0	10 −0	4 −1	10 −1	3 −2	9 −2
F	3 −0	0 +2	4 +4	5 −1	1 +4	6 −1
G	5 +0	1 +3	3 +5	7 +1	1 +7	4 +5

TR82 Teacher's Resource Book

DFP-1

Daily Facts Practice

A	2 +0	2 +1	3 +4	0 +6	6 +1	5 +3
B	6 +0	0 +1	5 +2	6 −0	1 −0	7 −2
C	3 −1	4 −2	6 −3	3 −3	9 −0	10 −2
D	2 −0	1 −1	5 −3	9 +0	7 +2	4 +3
E	7 +3	6 +4	8 +2	0 +10	5 +5	9 +1
F	3 +0	8 −1	3 +7	6 −4	0 +4	10 −3
G	1 +5	8 +3	5 +4	3 +6	3 +1	2 +7

DFP-2   Teacher's Resource Book   TR83

Daily Facts Practice

A	1 +0	1 +6	2 +2	3 +2	0 +9	4 +1
B	6 +2	9 −1	1 +10	7 −3	9 +3	9 −4
C	5 −2	6 +3	7 −0	8 +0	6 −2	5 +6
D	8 −4	5 −0	4 −3	8 −3	10 −4	11 −1
E	9 −3	10 −5	11 −0	4 −4	7 −5	8 −6
F	7 −7	7 −6	10 −9	5 −5	9 −6	8 −5
G	0 +5	11 −5	9 −5	6 −5	2 +5	7 +5

TR84 Teacher's Resource Book

DFP-3

Daily Facts Practice

A	11 −2	0 +8	9 −7	2 +9	6 +6	10 −7
B	9 −8	12 −0	5 −4	8 −8	13 −3	12 −5
C	4 +0	2 +10	12 −1	10 −8	11 −9	4 +8
D	9 −9	12 −3	1 +8	3 +9	4 +7	11 −3
E	10 +0	8 +4	2 +8	7 +4	5 +7	3 +10
F	4 +9	6 −6	7 +7	12 −9	3 +8	13 −6
G	12 −6	10 +1	11 −4	5 +8	16 −6	8 +7

Daily Facts Practice

A	10 +5	9 +4	6 +7	6 +9	4 +10	8 +5
B	12 −2	13 −4	8 −7	11 −8	10 −10	19 −9
C	7 +6	10 +3	8 +8	13 −7	12 −4	15 −10
D	12 −10	5 +10	14 −7	13 −5	8 +6	10 +4
E	6 +10	9 +5	10 +7	7 +8	8 +10	7 +9
F	14 −4	16 −7	13 −8	13 −9	16 −10	14 −5
G	7 +10	10 +6	9 +10	10 +8	9 +9	10 +10

TR86 Teacher's Resource Book

Daily Facts Practice

A	9 +6	5 +9	9 +7	10 +9	9 +8	8 +9
B	7 +4	0 +8	2 +5	4 +4	10 +3	6 +8
C	9 +4	7 +5	17 −10	16 −8	17 −9	14 −6
D	15 −9	19 −10	11 −6	14 −9	17 −7	16 −9
E	9 +9	7 +7	4 +4	5 +5	3 +3	8 +8
F	3 +10	4 +6	3 +1	5 +7	6 +9	9 +8
G	1 +5	10 +2	4 +8	3 +5	5 +9	6 +7

Daily Facts Practice

A	14 −10	10 −6	12 −8	18 −9	15 −6	20 −10

B	6 +5	18 −8	9 +2	11 −10	8 +6	12 −7

C	8 +3	7 +10	6 +6	0 +0	3 +7	7 +9

D	9 +1	7 +2	5 +3	2 +9	6 +4	5 +10

E	1 +3	5 +2	4 +7	10 +9	2 +6	8 +5

F	18 −10	15 −5	11 −7	14 −8	13 −10	15 −7

G	3 +8	5 +8	7 +8	17 −8	15 −8	13 −8

Daily Facts Practice

A	6 +1	5 +4	9 +5	10 +1	3 +9	5 +6

B	9 +0	10 +7	8 +2	2 +4	10 +10	8 +9

C	8 +10	3 +4	9 +7	10 +4	7 +6	6 +8

D	10 −9	8 −5	10 +6	2 +8	14 −6	17 −9

E	14 −10	8 −4	9 −5	12 −8	6 −2	13 −9

F	13 −7	9 −7	7 −7	3 −3	11 −10	9 −6

G	20 −10	12 −4	11 −8	2 −0	9 −3	15 −7

DFP-8   Teacher's Resource Book   TR89

Daily Facts Practice

A	12 −9	12 −3	3 +9	9 +3	4 +5	6 +0
B	6 +10	2 +2	11 −7	7 −6	16 −9	13 −6
C	5 −2	4 −0	7 −3	10 −7	19 −10	10 −3
D	16 −8	2 −1	5 −3	17 −8	10 −2	9 −4
E	15 −10	12 −6	9 −1	4 −4	14 −8	11 −9
F	12 −0	1 +6	10 −8	11 −6	2 +7	4 −2
G	3 +2	0 +6	8 +7	15 −6	6 −3	18 −8

TR90 Teacher's Resource Book

DFP-9

Daily Facts Practice

A	17 −10	14 −5	9 −8	6 −0	11 −5	16 −7
B	7 −1	8 −6	15 −9	13 −4	12 −7	14 −9
C	11 −2	6 −1	10 −5	2 −2	13 −10	7 −5
D	15 −5	10 −6	14 −7	6 +3	1 +8	4 +10
E	12 −10	13 −3	7 −2	12 −5	18 −9	11 −3
F	2 ×3	2 ×9	2 ×2	2 ×4	2 ×5	2 ×7
G	5 ×0	5 ×3	5 ×2	5 ×6	5 ×8	5 ×4

Daily Facts Practice

A	2 ×1	5 ×7	2 ×8	2 ×0	5 ×5	5 ×9
B	1 +1	8 +4	10 +5	4 +0	9 +6	7 +1
C	2 ×6	5 ×1	3 ×8	2 ×10	3 ×1	3 ×6
D	5 ×10	3 ×7	3 ×9	3 ×3	3 ×5	3 ×0
E	3 ×10	0 ×7	3 ×4	0 ×0	3 ×2	0 ×10
F	4 +1	19 −9	3 +6	13 −5	10 +8	15 −8
G	1 ×1	0 ×6	1 ×6	0 ×4	0 ×9	1 ×0

TR92 Teacher's Resource Book   DFP-11

Daily Facts Practice

A	0 ×1	1 ×7	4 ×2	4 ×4	0 ×3	1 ×9
B	4 ×1	0 ×8	1 ×10	0 ×2	1 ×3	4 ×7
C	0 +10	1 +7	4 +9	2 +10	7 +3	6 +2
D	1 ×5	4 ×6	4 ×0	0 ×5	1 ×8	4 ×5
E	1 ×4	6 ×3	4 ×10	6 ×5	6 ×4	4 ×8
F	6 ×6	4 ×3	6 ×0	7 ×5	6 ×8	4 ×9
G	10 −10	0 +1	11 −4	2 +3	6 −4	9 +7

DFP-12

Daily Facts Practice

A	2)¯18	2)¯12	2)¯0	5)¯5	5)¯30	5)¯50
B	5)¯10	2)¯4	5)¯45	2)¯8	5)¯25	2)¯10
C	2)¯14	2)¯2	5)¯0	5)¯15	3)¯30	3)¯15
D	6 ×10	7 ×2	6 ×2	2)¯6	5)¯35	3)¯9
E	5)¯20	2)¯16	3)¯21	6 ×1	7 ×4	7 ×9
F	9 −9	8 −2	18 −10	7 −4	10 −4	15 −6
G	7 ×0	6 ×9	6 ×7	7 ×10	8 ×7	7 ×3

TR94 Teacher's Resource Book

Daily Facts Practice

A	7 ×6	5)40	8 ×1	2)20	8 ×8	3)24
B	3)0	3)12	4)12	3)6	3)27	4)20
C	4)8	3)3	4)28	3)18	4)40	4)32
D	7 ×1	8 ×5	8 ×10	7 ×8	8 ×2	7 ×7
E	4)0	8 ×4	1)4	8 ×0	1)10	8 ×6
F	4)4	1)6	4)24	4)36	1)9	4)32
G	9 ×3	9 ×5	6)24	6)6	9 ×0	6)30

Teacher's Resource Book   TR95

Daily Facts Practice

A	8 ×3	8 ×9	9 ×10	9 ×2	9 ×6	9 ×7
B	1)1	10 ×2	6)0	1)2	9 ×9	9 ×4
C	1 +9	4 +2	3 +6	8 +0	6 +9	8 +4
D	16 −10	8 −7	8 −3	9 −2	12 −5	14 −7
E	10 ×0	10 ×6	9 ×8	10 ×5	10 ×10	9 ×1
F	1)7	6)60	6)48	1)0	6)12	6)54
G	10 ×9	10 ×7	6)18	1)8	10 ×1	6)36

TR96 Teacher's Resource Book   DFP-15

Daily Facts Practice

A	$6\overline{)42}$	$1\overline{)3}$	$\begin{array}{r}10\\ \times 4\\ \hline\end{array}$	$\begin{array}{r}10\\ \times 3\\ \hline\end{array}$	$1\overline{)5}$	$\begin{array}{r}10\\ \times 8\\ \hline\end{array}$
B	$\begin{array}{r}8\\ \times 4\\ \hline\end{array}$	$\begin{array}{r}10\\ \times 2\\ \hline\end{array}$	$\begin{array}{r}6\\ \times 6\\ \hline\end{array}$	$\begin{array}{r}1\\ \times 7\\ \hline\end{array}$	$\begin{array}{r}4\\ \times 6\\ \hline\end{array}$	$\begin{array}{r}9\\ \times 8\\ \hline\end{array}$
C	$7\overline{)21}$	$8\overline{)40}$	$7\overline{)0}$	$8\overline{)16}$	$7\overline{)49}$	$8\overline{)64}$
D	$\begin{array}{r}7\\ \times 3\\ \hline\end{array}$	$8\overline{)0}$	$7\overline{)14}$	$8\overline{)48}$	$7\overline{)42}$	$\begin{array}{r}4\\ \times 8\\ \hline\end{array}$
E	$\begin{array}{r}4\\ \times 9\\ \hline\end{array}$	$\begin{array}{r}10\\ \times 5\\ \hline\end{array}$	$\begin{array}{r}2\\ \times 6\\ \hline\end{array}$	$\begin{array}{r}5\\ \times 7\\ \hline\end{array}$	$\begin{array}{r}7\\ \times 7\\ \hline\end{array}$	$\begin{array}{r}9\\ \times 6\\ \hline\end{array}$
F	$\begin{array}{r}4\\ \times 0\\ \hline\end{array}$	$\begin{array}{r}3\\ \times 2\\ \hline\end{array}$	$\begin{array}{r}5\\ \times 4\\ \hline\end{array}$	$\begin{array}{r}10\\ \times 7\\ \hline\end{array}$	$\begin{array}{r}7\\ \times 1\\ \hline\end{array}$	$\begin{array}{r}6\\ \times 8\\ \hline\end{array}$
G	$\begin{array}{r}5\\ \times 6\\ \hline\end{array}$	$\begin{array}{r}8\\ \times 2\\ \hline\end{array}$	$\begin{array}{r}6\\ \times 3\\ \hline\end{array}$	$\begin{array}{r}8\\ \times 8\\ \hline\end{array}$	$\begin{array}{r}7\\ \times 9\\ \hline\end{array}$	$\begin{array}{r}3\\ \times 9\\ \hline\end{array}$

DFP-16

Daily Facts Practice

A	4 × 3	7)7	7)63	8)24	8)72	8 × 5
B	3 × 5	8)32	8)8	7)35	7)56	9 × 4
C	9 × 7	7)70	8)48	9)90	9)18	4 × 7
D	9)36	8)80	7)28	9)9	9)45	10)50
E	10)90	10)60	9)0	9)63	10)10	9)81
F	6 × 2	10)30	9)27	3 × 10	0 × 5	10)80
G	1 × 2	4 × 4	10)0	9)72	10)100	8 × 9

TR98 Teacher's Resource Book   DFP-17

Daily Facts Practice

A	$\begin{array}{r}6\\ \times 10\\ \hline\end{array}$	$10\overline{)20}$	$\begin{array}{r}5\\ \times 3\\ \hline\end{array}$	$10\overline{)40}$	$9\overline{)54}$	$10\overline{)70}$
B	$\begin{array}{r}3\\ \times 7\\ \hline\end{array}$	$\begin{array}{r}2\\ \times 9\\ \hline\end{array}$	$\begin{array}{r}5\\ \times 8\\ \hline\end{array}$	$\begin{array}{r}3\\ \times 3\\ \hline\end{array}$	$\begin{array}{r}6\\ \times 4\\ \hline\end{array}$	$\begin{array}{r}2\\ \times 7\\ \hline\end{array}$
C	$\begin{array}{r}10\\ \times 10\\ \hline\end{array}$	$\begin{array}{r}0\\ \times 6\\ \hline\end{array}$	$\begin{array}{r}7\\ \times 8\\ \hline\end{array}$	$\begin{array}{r}1\\ \times 3\\ \hline\end{array}$	$\begin{array}{r}9\\ \times 5\\ \hline\end{array}$	$\begin{array}{r}9\\ \times 2\\ \hline\end{array}$
D	$\begin{array}{r}7\\ \times 5\\ \hline\end{array}$	$5\overline{)35}$	$\begin{array}{r}1\\ \times 6\\ \hline\end{array}$	$6\overline{)6}$	$\begin{array}{r}6\\ \times 9\\ \hline\end{array}$	$9\overline{)54}$
E	$\begin{array}{r}4\\ +3\\ \hline\end{array}$	$\begin{array}{r}7\\ +9\\ \hline\end{array}$	$\begin{array}{r}6\\ +5\\ \hline\end{array}$	$\begin{array}{r}2\\ +7\\ \hline\end{array}$	$\begin{array}{r}4\\ +8\\ \hline\end{array}$	$\begin{array}{r}8\\ +3\\ \hline\end{array}$
F	$\begin{array}{r}9\\ +10\\ \hline\end{array}$	$\begin{array}{r}5\\ +3\\ \hline\end{array}$	$\begin{array}{r}8\\ +6\\ \hline\end{array}$	$\begin{array}{r}2\\ +4\\ \hline\end{array}$	$\begin{array}{r}5\\ +6\\ \hline\end{array}$	$\begin{array}{r}4\\ +7\\ \hline\end{array}$
G	$\begin{array}{r}3\\ -2\\ \hline\end{array}$	$\begin{array}{r}10\\ -4\\ \hline\end{array}$	$\begin{array}{r}5\\ -5\\ \hline\end{array}$	$\begin{array}{r}13\\ -9\\ \hline\end{array}$	$\begin{array}{r}17\\ -7\\ \hline\end{array}$	$\begin{array}{r}12\\ -9\\ \hline\end{array}$

DFP-18

Daily Facts Practice

A	14 −8	11 −7	12 −3	6 −6	8 −5	9 −8
B	10 ×4	3 ×8	4 ×5	7 ×2	2 ×2	7 ×4
C	6 ×5	2 ×8	7 ×6	9 ×3	10 ×9	3 ×4
D	2 ×3	8 ×6	7 ×10	4)20	6)18	9)45
E	4)4	8)72	7)63	5 ×9	4 ×2	10 ×3
F	9)81	4)16	6)36	7)49	1)1	8)64
G	10)70	6)48	2)12	5)5	7)28	6)24

Daily Facts Practice

A	$\begin{array}{r}5\\\times 2\\\hline\end{array}$	$4\overline{)40}$	$\begin{array}{r}9\\\times 9\\\hline\end{array}$	$2\overline{)18}$	$\begin{array}{r}8\\\times 3\\\hline\end{array}$	$5\overline{)30}$
B	$\begin{array}{r}2\\\times 4\\\hline\end{array}$	$\begin{array}{r}10\\\times 6\\\hline\end{array}$	$\begin{array}{r}5\\\times 5\\\hline\end{array}$	$\begin{array}{r}3\\\times 6\\\hline\end{array}$	$\begin{array}{r}8\\\times 7\\\hline\end{array}$	$\begin{array}{r}6\\\times 7\\\hline\end{array}$
C	$\begin{array}{r}2\\\times 10\\\hline\end{array}$	$\begin{array}{r}10\\\times 5\\\hline\end{array}$	$\begin{array}{r}4\\\times 10\\\hline\end{array}$	$\begin{array}{r}10\\\times 8\\\hline\end{array}$	$\begin{array}{r}2\\\times 5\\\hline\end{array}$	$\begin{array}{r}5\\\times 5\\\hline\end{array}$
D	$\begin{array}{r}10\\+10\\\hline\end{array}$	$\begin{array}{r}10\\-5\\\hline\end{array}$	$\begin{array}{r}5\\+5\\\hline\end{array}$	$\begin{array}{r}15\\-5\\\hline\end{array}$	$\begin{array}{r}10\\+5\\\hline\end{array}$	$\begin{array}{r}20\\-10\\\hline\end{array}$
E	$\begin{array}{r}4\\-3\\\hline\end{array}$	$\begin{array}{r}16\\-6\\\hline\end{array}$	$\begin{array}{r}15\\-8\\\hline\end{array}$	$\begin{array}{r}11\\-9\\\hline\end{array}$	$\begin{array}{r}6\\-3\\\hline\end{array}$	$\begin{array}{r}9\\-5\\\hline\end{array}$
F	$\begin{array}{r}7\\\times 4\\\hline\end{array}$	$\begin{array}{r}6\\\times 5\\\hline\end{array}$	$\begin{array}{r}9\\\times 5\\\hline\end{array}$	$\begin{array}{r}2\\\times 8\\\hline\end{array}$	$\begin{array}{r}7\\\times 7\\\hline\end{array}$	$\begin{array}{r}3\\\times 4\\\hline\end{array}$
G	$\begin{array}{r}2\\\times 2\\\hline\end{array}$	$\begin{array}{r}5\\\times 6\\\hline\end{array}$	$\begin{array}{r}9\\\times 8\\\hline\end{array}$	$\begin{array}{r}3\\\times 6\\\hline\end{array}$	$\begin{array}{r}4\\\times 8\\\hline\end{array}$	$\begin{array}{r}7\\\times 2\\\hline\end{array}$

Daily Facts Practice

A	3 ×3	4 ×7	9 ×9	6 ×8	4 ×3	8 ×7
B	8)56	3 ×9	4)28	4 ×5	1 ×8	10)20
C	10)50	7)21	4)36	3)15	6)48	7)56
D	1)6	8)80	2)8	6)54	9)18	5)40
E	3)9	6)30	3)24	7)42	10)80	9)36
F	4 ×2	9 ×4	3 ×7	8)24	5)15	9)72
G	6 ×9	5 ×7	7)14	8)40	8 ×8	9)27

TR102 Teacher's Resource Book DFP-21

Daily Facts Practice

A	6 ×6	2 ×4	9 ×3	2 ×7	7 ×8	5 ×9
B	8)16	3)21	9)63	6 ×3	7 ×6	3 ×8
C	2 ×6	4 ×9	5 ×3	6 ×4	9 ×6	2 ×5
D	8 ×2	7 ×5	2 ×3	6 ×7	4 ×6	8 ×9
E	3 ×5	8)32	5 ×4	5)10	8 ×3	4)24
F	2)14	9 ×2	8 ×6	7 ×3	3 ×2	7)35
G	8 ×5	3)12	2 ×9	4 ×4	4)32	7 ×9

DFP-22

Teacher's Resource Book   TR103

Daily Facts Practice

A	2)6̄	6)4̄2̄	5)4̄5̄	6)1̄2̄	3)2̄7̄	3)1̄8̄
B	5 ×2	8 ×4	5 ×8	9 ×7	7 ×7	5 ×6
C	6 ×2	5)2̄0̄	4)1̄2̄	2)1̄6̄	9)2̄7̄	6 ×8
D	5)2̄5̄	8)2̄4̄	7)4̄9̄	8)6̄4̄	9)7̄2̄	10)1̄0̄0̄
E	7 ×9	2)4̄	4 ×8	4)8̄	8 ×8	6)4̄2̄

Fact Cards

0 × 0	1 × 0	2 × 0
3 × 0	4 × 0	5 × 0
6 × 0	7 × 0	8 × 0

Teacher's Resource Book TR105

$\begin{array}{r}9\\\times 0\\\hline\end{array}$	$\begin{array}{r}10\\\times 0\\\hline\end{array}$	$\begin{array}{r}0\\\times 1\\\hline\end{array}$
$\begin{array}{r}1\\\times 1\\\hline\end{array}$	$\begin{array}{r}2\\\times 1\\\hline\end{array}$	$\begin{array}{r}3\\\times 1\\\hline\end{array}$
$\begin{array}{r}4\\\times 1\\\hline\end{array}$	$\begin{array}{r}5\\\times 1\\\hline\end{array}$	$\begin{array}{r}6\\\times 1\\\hline\end{array}$

TR106   Teacher's Resource Book   Fact Cards

7 × 1	8 × 1	9 × 1
10 × 1	0 × 2	1 × 2
2 × 2	3 × 2	4 × 2

Fact Cards Teacher's Resource Book **TR 107**

$\begin{array}{r}5\\ \times 2\\ \hline\end{array}$	$\begin{array}{r}6\\ \times 2\\ \hline\end{array}$	$\begin{array}{r}7\\ \times 2\\ \hline\end{array}$
$\begin{array}{r}8\\ \times 2\\ \hline\end{array}$	$\begin{array}{r}9\\ \times 2\\ \hline\end{array}$	$\begin{array}{r}10\\ \times 2\\ \hline\end{array}$
$\begin{array}{r}0\\ \times 3\\ \hline\end{array}$	$\begin{array}{r}1\\ \times 3\\ \hline\end{array}$	$\begin{array}{r}2\\ \times 3\\ \hline\end{array}$

$\begin{array}{r} 3 \\ \times 3 \\ \hline \end{array}$	$\begin{array}{r} 4 \\ \times 3 \\ \hline \end{array}$	$\begin{array}{r} 5 \\ \times 3 \\ \hline \end{array}$
$\begin{array}{r} 6 \\ \times 3 \\ \hline \end{array}$	$\begin{array}{r} 7 \\ \times 3 \\ \hline \end{array}$	$\begin{array}{r} 8 \\ \times 3 \\ \hline \end{array}$
$\begin{array}{r} 9 \\ \times 3 \\ \hline \end{array}$	$\begin{array}{r} 10 \\ \times 3 \\ \hline \end{array}$	$\begin{array}{r} 0 \\ \times 4 \\ \hline \end{array}$

Fact Cards — Teacher's Resource Book TR109

1	2	3
×4	×4	×4

4	5	6
×4	×4	×4

7	8	9
×4	×4	×4

TR110   **Teacher's Resource Book**   Fact Cards

Fact Cards

10 ×4	0 ×5	1 ×5
2 ×5	3 ×5	4 ×5
5 ×5	6 ×5	7 ×5

Teacher's Resource Book **TR 111**

$\begin{array}{r}8\\ \times 5\\ \hline\end{array}$	$\begin{array}{r}9\\ \times 5\\ \hline\end{array}$	$\begin{array}{r}10\\ \times 5\\ \hline\end{array}$
$\begin{array}{r}0\\ \times 6\\ \hline\end{array}$	$\begin{array}{r}1\\ \times 6\\ \hline\end{array}$	$\begin{array}{r}2\\ \times 6\\ \hline\end{array}$
$\begin{array}{r}3\\ \times 6\\ \hline\end{array}$	$\begin{array}{r}4\\ \times 6\\ \hline\end{array}$	$\begin{array}{r}5\\ \times 6\\ \hline\end{array}$

Fact Cards

6 × 6	7 × 6	8 × 6
9 × 6	10 × 6	0 × 7
1 × 7	2 × 7	3 × 7

Teacher's Resource Book   TR113

TR114   Teacher's Resource Book

| 4 | 5 | 6 |
| ×7 | ×7 | ×7 |

| 7 | 8 | 9 |
| ×7 | ×7 | ×7 |

| 10 | 0 | 1 |
| ×7 | ×8 | ×8 |

Fact Cards

2 ×8	3 ×8	4 ×8
5 ×8	6 ×8	7 ×8
8 ×8	9 ×8	10 ×8

Fact Cards

Teacher's Resource Book **TR 115**

0	1	2
×9	×9	×9

3	4	5
×9	×9	×9

6	7	8
×9	×9	×9

9 ×9	10 ×9	0 ×10
1 ×10	2 ×10	3 ×10
4 ×10	5 ×10	6 ×10

Fact Cards   Teacher's Resource Book   TR117

| 7 × 10 | 8 × 10 | 9 × 10 |

10 × 10

TR118 Teacher's Resource Book

Fact Cards

$1\overline{)2}$	$1\overline{)5}$	$1\overline{)8}$
$1\overline{)1}$	$1\overline{)4}$	$1\overline{)7}$
$1\overline{)0}$	$1\overline{)3}$	$1\overline{)6}$

Fact Cards Teacher's Resource Book **TR 119**

2)0	2)6	2)12
1)10	2)4	2)10
1)9	2)2	2)8

TR120 Teacher's Resource Book

Fact Cards

$2\overline{)18}$	$3\overline{)3}$	$3\overline{)12}$
$2\overline{)16}$	$3\overline{)0}$	$3\overline{)9}$
$2\overline{)14}$	$2\overline{)20}$	$3\overline{)6}$

Fact Cards

Teacher's Resource Book TR121

3)21	3)30	4)8
3)18	3)27	4)4
3)15	3)24	4)0

TR122 Teacher's Resource Book

Fact Cards

$4\overline{)20}$	$4\overline{)32}$	$5\overline{)0}$
$4\overline{)16}$	$4\overline{)28}$	$4\overline{)40}$
$4\overline{)12}$	$4\overline{)24}$	$4\overline{)36}$

Fact Cards

$5\overline{)15}$	$5\overline{)30}$	$5\overline{)45}$
$5\overline{)10}$	$5\overline{)25}$	$5\overline{)40}$
$5\overline{)5}$	$5\overline{)20}$	$5\overline{)35}$

TR124   Teacher's Resource Book   Fact Cards

$6\overline{)6}$	$6\overline{)24}$	$6\overline{)42}$
$6\overline{)0}$	$6\overline{)18}$	$6\overline{)36}$
$5\overline{)50}$	$6\overline{)12}$	$6\overline{)30}$

Fact Cards Teacher's Resource Book TR125

$6\overline{)60}$	$7\overline{)14}$	$7\overline{)35}$
$6\overline{)54}$	$7\overline{)7}$	$7\overline{)28}$
$6\overline{)48}$	$7\overline{)0}$	$7\overline{)21}$

$7\overline{)56}$	$8\overline{)0}$	$8\overline{)24}$
$7\overline{)49}$	$7\overline{)70}$	$8\overline{)16}$
$7\overline{)42}$	$7\overline{)63}$	$8\overline{)8}$

Fact Cards · Teacher's Resource Book · TR127

8)48	8)72	9)9
8)40	8)64	9)0
8)32	8)56	8)80

TR128 Teacher's Resource Book

Fact Cards

Fact Cards

9)36	9)63	9)90
9)27	9)54	9)81
9)18	9)45	9)72

Teacher's Resource Book TR129

$10\overline{)0}$	$10\overline{)10}$	$10\overline{)20}$
$10\overline{)30}$	$10\overline{)40}$	$10\overline{)50}$
$10\overline{)60}$	$10\overline{)70}$	$10\overline{)80}$

$10\overline{)100}$

$10\overline{)90}$

Fact Cards

Teacher's Resource Book TR 131

VOCABULARY CARDS

Use the vocabulary cards to practice and review this year's new math terms. Suggestions for using the cards are in the Teacher's Edition, on the Chapter at a Glance page.

Consider having students organize their vocabulary cards in Math Words Files—containers made from zip-top bags or small boxes, such as crayon or computer disk boxes. Encourage students to consult their Math Word Files to confirm meanings, verify pronunciations, and check spellings.

To copy the cards, set the copy machine to 2-sided copies. Align the perforated edge with the left-hand (or top) guide on the glass and copy. Flip the page and align the perforated edge with the opposite (right-hand or bottom) guide. Copy.

Pronunciation Key

a	add, map	h	hope, hate	ô	order, jaw	th	this, bathe
ā	ace, rate	i	it, give	oi	oil, boy	u	up, done
â(r)	care, air	ī	ice, write	ou	pout, now	û(r)	burn, term
ä	palm, father	j	joy, ledge	o͝o	took, full	yo͞o	fuse, few
b	bat, rub	k	cool, take	o͞o	pool, food	v	vain, eve
ch	check, catch	l	look, rule	p	pit, stop	w	win, away
d	dog, rod	m	move, seem	r	run, poor	y	yet, yearn
e	end, pet	n	nice, tin	s	see, pass	z	zest, muse
ē	equal, tree	ng	ring, song	sh	sure, rush	zh	vision, pleasure
f	fit, half	o	odd, hot	t	talk, sit		
g	go, log	ō	open, so	th	thin, both		

ə the schwa, an unstressed vowel representing the sound spelled *a* in **a**bove, *e* in sick**e**n, *i* in poss**i**ble, *o* in mel**o**n, *u* in circ**u**s

Other symbols:
• separates words into syllables
´ indicates stress on a syllable

even	odd
digits	expanded form
standard form	word form
benchmark numbers	compare

Vocabulary Cards

od **A whole number that has a 1, 3, 5, 7, or 9 in the ones place is odd. (2)**	ē′vən **A whole number that has a 0, 2, 4, 6, or 8 in the ones place is even. (2)**
ik•spand′id fôrm′ **A way to write numbers by showing the value of each digit (4)**	di′jəts **The symbols 0, 1, 2, 3, 4, 5, 6, 7, 8, and 9 (4)**
wûrd form **A way to write numbers by using words (4)**	stan′dərd fôrm′ **A way to write numbers by using the digits 0–9, with each digit having a place value (4)**
kəm•par′ **To describe whether numbers are equal to, less than, or greater than each other (20)**	bench′märk num′bərz **Numbers that help you estimate the number of objects without counting them, such as 25, 50, 100, 1,000 (18)**

equal sign (=)	greater than (>)
less than (<)	rounding
Grouping Property of Addition	estimate
expression	equivalent

Vocabulary Cards

grā′tər ŧħan **A symbol used to compare two numbers, with the greater number given first (20)**	ē′kwəl sīn **A symbol used to show that two numbers have the same value (20)**
roun′ding **One way to estimate (28)**	les ŧħan **A symbol used to compare two numbers, with the lesser number given first (20)**
es′tə•māt **To find about how many or how much (38)**	groo′ping prä′pər•tē əv ə•di′shən **A rule stating that you can group addends in different ways and still get the same sum (36)**
ē•kwiv′ə•lənt **Two or more sets that name the same amount are equivalent. (80)**	ik•spre′shən **The part of a number sentence that combines numbers and operation signs, but doesn't have an equal sign (68)**

TR138 Teacher's Resource Book

Vocabulary Cards

minute (min)	A.M.
midnight	noon
P.M.	elapsed time
schedule	calendar

ā em **Between midnight and noon (96)**	mi′nət **A unit used to measure short amounts of time; in one minute, the minute hand moves from one mark to the next (94)**
no͞on **12:00 in the day (96)**	mid′nīt **12:00 at night (96)**
i•lapst′ tīm **The amount of time that passes from the start of an activity to the end of that activity (98)**	pē em **Between noon and midnight (96)**
ka′lən•dər **A table that shows the days, weeks, and months of a year (102)**	ske′jo͞ol **A table that lists activities or events and the times they happen (100)**

multiply	factor
product	array
Order Property of Multiplication	Grouping Property of Multiplication
multistep problem	divide

fak′tər A number that is multiplied by another number to find a product (118)	**mul′tə•plī** When you combine equal groups, you can multiply to find how many in all; the opposite operation of division (116)
ə•rā′ An arrangement of objects in rows and columns (120)	**prä′dəkt** The answer in a multiplication problem (118)
gro͞o′ping prä′pər•tē əv məl•tə•plə•kā′shən A rule stating that when the grouping of factors is changed, the product remains the same (170)	**ôr′dər prä′pər•tē əv məl•tə•plə•kā′shən** A rule stating that you can multiply two factors in any order and get the same product (121)
di•vīd′ To separate into equal groups; the opposite operation of multiplication (184)	**mul′tē•step prä′bləm** A problem with more than one step (172)

dividend	divisor
inverse operations	quotient
variable	fact family
data	frequency table

də•vī′zər **The number that divides the dividend (188)**	di′və•dend **The number that is to be divided in a division problem (188)**
kwō′shənt **The number, not including the remainder, that results from dividing (188)**	in′vərs ä•pə•rā′shənz **Opposite operations, or operations that undo each other, such as addition and subtraction or multiplication and division (188)**
fakt fam′ə•lē **A set of related multiplication and division, or addition and subtraction, number sentences (192)**	vâr′ē•ə•bəl **A symbol or a letter that stands for an unknown number (189)**
frē′kwen•sē tā′bəl **A table that uses numbers to record data (240)**	dā′tə **Information collected about people or things (240)**

tally table	survey
results	classify
bar graph	horizontal bar graph
scale	vertical bar graph

Vocabulary Cards

Teacher's Resource Book **TR145**

sər′vā **A question or set of questions that a group of people are asked (242)**	ta′lē tā′bəl **A table that uses tally marks to record data (240)**
kla′sə•fī **To group pieces of data according to how they are the same; for example, you can classify data by size, color, or shape (244)**	ri•zults′ **The answers from a survey (242)**
hôr•ə•zän′təl bär graf **A bar graph in which the bars go from left to right (254)**	bär graf **A graph that uses bars to show data (254)**
vûr′ti•kəl bär graf **A bar graph in which the bars go up from bottom to top (254)**	skāl **The numbers on a bar graph that help you read the number each bar shows (254)**

mode	range
line plot	grid
ordered pair	line graph
certain	impossible

Vocabulary Cards   Teacher's Resource Book **TR147**

rānj **The difference between the greatest number and the least number in a set of data (258)**	mōd **The number found most often in a set of data (258)**
grid **Horizontal and vertical lines on a map (262)**	līn plot **A diagram that records each piece of data on a number line (258)**
līn graf **A graph that uses a line to show how something changes over time (264)**	ôr′dərd pâr′ **A pair of numbers that names a point on a grid (262)**
im•pä′sə•bəl **An event is impossible if it will never happen. (270)**	sûr′tən **An event is certain if it will always happen. (270)**

Teacher's Resource Book Vocabulary Cards

event	likely
unlikely	outcome
equally likely	possible outcome
predict	fair

Vocabulary Cards — Teacher's Resource Book **TR149**

lĭk′lē **Having a good chance of happening (272)**	i•vent′ **Something that happens (270)**
out′kum′ **A possible result of an experiment (272)**	ən•lī′klē **An event is unlikely if it does not have a good chance of happening (272)**
pos′ə•bəl out′kəm **Something that has a chance of happening (274)**	ē′kwəl•lē lī′klē **Having the same chance of happening (274)**
fâr **A game is fair if every player has an equal chance to win. (282)**	pri•dikt′ **To make a reasonable guess about what will happen (274)**

face	edge
vertex	point
right angle	line
line segment	ray

ej **A line segment formed where two faces meet (294)**	fās **A flat surface of a solid figure (294)**
point **An exact position or location (300)**	vûr′teks **In a solid figure, a corner where three or more edges meet (294)**
līn **A straight path extending in both directions with no endpoints (300)**	rīt ang′gəl **A special angle that forms a square corner (300)**
rā **A part of a line, with one endpoint, that is straight and continues in one direction (301)**	līn seg′mənt **A part of a line that extends between two points, called endpoints (300)**

angle	intersecting lines
parallel lines	center
diameter	radius
hexagon	octagon

Vocabulary Cards Teacher's Resource Book TR153

in•tər•sek′ting līnz **Lines that cross (304)**	ang′gəl **The figure formed when two rays share the same endpoint (301)**
sen′tər **A point in the middle of a circle that is the same distance from any where on the circle (307)**	par′ə•lel līnz **Lines that never cross (304)**
rā′dē•əs **A line segment whose endpoints are the center of a circle and any point on the circle (307)**	dī•a′mə•tər **A line segment that passes through the center of a circle and whose endpoints are on the circle (307)**
äk′tə•gän **A polygon with eight sides and eight angles (314)**	hek′sə•gän **A polygon with six sides and six angles (314)**

pentagon	polygon
quadrilateral	right triangle
equilateral triangle	isosceles triangle
scalene triangle	acute triangle

pol′ē•gän **A closed plane figure with straight sides; each side is a line segment. (314)**	pen′tə•gän **A polygon with five sides and five angles (314)**
rīt trī′ang•gəl **A triangle with one right angle (317)**	kwa•drə•lat′ər•əl **A polygon with four sides and four angles (314)**
ī•sas′ə•lēz trī′ang•gəl **A triangle that has two equal sides (317)**	ē•kwə•lat′ər•əl trī′ang•gəl **A triangle with three equal sides (317)**
ə•kyo͞ot′ trī′ang•gəl **A triangle with three angles less than a right angle (317)**	skā′lēn trī′ang•gəl **A triangle in which no sides are equal (317)**

obtuse triangle	parallelogram
rhombus	tessellate
tessellation	congruent
line of symmetry	similar

par•ə•lel′ə•gram **A quadrilateral with 2 pairs of parallel sides and 2 pairs of equal sides (321)**	əb•tōōs′ trī′ang•gəl **A triangle with one angle greater than a right angle (317)**
tes′ə•lāt **To combine plane figures so they cover a surface without overlapping or leaving any space between them (324)**	räm′bəs **A quadrilateral with 2 pairs of parallel sides and 4 equal sides (321)**
kən•grōō′ənt **Having the same size and shape (332)**	te•sə•lā′shən **A repeating pattern of closed figures that covers a surface with no gaps and no overlaps (324)**
si′mə•lər **Having the same shape and the same or different size (336)**	līn əv si′mə•trē **An imaginary line that divides a figure into two congruent parts (334)**

slide	flip
turn	inch (in.)
yard (yd)	mile (mi)
foot (ft)	capacity

flip **To turn a plane figure over (339)**	slīd **To move a plane figure in one direction (339)**
inch **A customary unit used to measure length (352)**	turn **To rotate a plane figure (339)**
mīl **A customary unit used to measure length and distance; 1 mile = 5,280 feet (356)**	yärd **A customary unit used to measure length or distance; 1 yard = 3 feet (356)**
kə•pa′sə•tē **The amount a container can hold (358)**	fo͝ot **A customary unit used to measure length or distance; 1 foot = 12 inches (356)**

cup (c)	gallon (gal)
quart (qt)	pint (pt)
ounce (oz)	pound (lb)
decimeter (dm)	centimeter (cm)

ga′lən **A customary unit for measuring capacity; 4 quarts = 1 gallon (359)**	kup **A customary unit used to measure capacity (358)**
pīnt **A customary unit for measuring capacity; 1 pint = 2 cups (358)**	kwôrt **A customary unit for measuring capacity; 1 quart = 2 pints (358)**
pound **A customary unit used to measure weight; 1 pound = 16 ounces (360)**	ouns **A customary unit used to measure weight (360)**
sən′tə•mē•tər **A metric unit that is used to measure length (372)**	de′sə•mē•tər **A metric unit that is used to measure length; 1 decimeter = 10 centimeters (372)**

meter (m)	kilometer (km)
liter (L)	milliliter (mL)
gram (g)	kilogram (kg)
degrees Celsius (°C)	degrees Fahrenheit (°F)

kə•lä′mə•tər **A metric unit that is used to measure length and distance; 1 kilometer = 1,000 meters (372)**	mē′tər **A metric unit that is used to measure length and distance; 1 meter = 100 centimeters (372)**
mi′lə•lē•tər **A metric unit that is used to measure capacity; 1 liter = 1,000 milliliters (378)**	lē′tər **A metric unit that is used to measure capacity; 1 liter = 1,000 milliliters (378)**
kil′ə•gram **A metric unit that is used to measure mass; 1 kilogram = 1,000 grams (380)**	gram **A metric unit that is used to measure mass (380)**
di•grēz′ far′ən•hīt **A unit for measuring temperature in the customary system (382)**	di•grēz′ səl′sē•əs **A unit for measuring temperature in the metric system (382)**

perimeter	area
square unit	cubic unit
volume	denominator
numerator	fraction

Vocabulary Cards

Teacher's Resource Book TR165

âr′ē•ə **The number of square units needed to cover a flat surface (394)**	pə•ri′mə•tər **The distance around a figure (388)**
kyoo′bik yoo′nət **A cube with a side length of one unit; used to measure volume (398)**	skwâr yoo′nət **A square with a side length of one unit; used to measure area (394)**
di•nä′mə•nā•tər **The part of a fraction that tells how many equal parts are in the whole (412)**	väl′yəm **The amount of space a solid figure takes up (398)**
frak′shən **A number that names part of a whole or part of a group (412)**	noo′mə•rā•tər **The part of a fraction above the line, which tells how many parts are being counted (412)**

TR166 Teacher's Resource Book   Vocabulary Cards

equivalent fractions	mixed number
like fractions	simplest form
decimal	tenth
hundredth	remainder

mikst nəm′bər **A number represented by a whole number and a fraction (428)**	ē•kwiv′ə•lənt frak′shənz **Two or more fractions that name the same amount (418)**
sim′pləst fôrm **When a fraction can be modeled with the largest fraction bar or bars possible (436)**	līk frak′shənz **Fractions that have the same denominator (434)**
tenth **One of ten equal parts (452)**	de′sə•məl **A number with one or more digits to the right of the decimal point (452)**
ri•mān′dər **The amount left over when a number cannot be divided evenly (503)**	hən′drədth **One of one hundred equal parts (456)**

TR168 Teacher's Resource Book

Vocabulary Cards